MATH MAKERS

The Lives and Works of 50 Famous Mathematicians

Alfred S. Posamentier
and Christian Spreitzer

Prometheus
Books

Guilford, Connecticut

Prometheus Books

An imprint of The Rowman & Littlefield Publishing Group, Inc.
4501 Forbes Blvd., Ste. 200
Lanham, MD 20706
www.rowman.com

Distributed by NATIONAL BOOK NETWORK

British Library Cataloguing in Publication Information available

Library of Congress Cataloging-in-Publication Data

Names: Posamentier, Alfred S., author. | Spreitzer, Christian, 1979– author.
Title: Math makers: the lives and works of 50 famous mathematicians / Alfred S. Posamentier and Christian Spreitzer.
Description: Amherst, New York: Prometheus Books, 2019. | Includes index.
Identifiers: LCCN 2018052509 (print) | LCCN 2018059061 (ebook) | ISBN 9781633885219 (ebook) | ISBN 9781633885202 (hardcover)
Subjects: LCSH: Mathematicians—Biography. | Mathematics—History.
Classification: LCC QA28 (ebook) | LCC QA28 .P67 2019 (print) | DDC 510.92/2—dc23
LC record available at https://lccn.loc.gov/2018052509

∞™ The paper used in this publication meets the minimum requirements of American National Standard for Information Sciences—Permanence of Paper for Printed Library Materials, ANSI/NISO Z39.48-1992

To my children and grandchildren, whose future is unbounded,
Lisa, Daniel, David, Lauren, Max, Samuel, Jack, and Charles
—Alfred S. Posamentier

To my mathematics instructors and mentors for
fostering my love for mathematics
—Christian Spreitzer

Contents

Introduction

It is unfortunate that in our culture, mathematics, in general, is viewed unfavorably by a majority of well-educated people. Oftentimes they are proud to admit that they were not good students of mathematics in their school days. Admitting this weakness is almost like a badge of honor, which is rarely attributed to any other school subject. Despite its negative reputation, mathematics has contributed immensely to humanity's shared knowledge of how the world works, and to the technological progress, which provides our lives with previously unforeseen advantages. Galileo Galilei stated that the book of nature is written in the language of mathematics, and, indeed, our understanding of nature through physics and other natural sciences is largely dependent on mathematics.[1] However, both the formal system of mathematics and all of the mathematical results that have been achieved to the present day are often seen as independent of the world around us. In principle, the mathematical knowledge we have was primarily developed without any interaction with nature at all. Unlike biology, for instance, mathematics is not an empirical science. Part of what makes mathematics a truly fascinating subject is that it is the universal language of nature, but—at the same time—it is a system of logical conclusions that can be continuously developed in the absence of any observations of natural phenomena. These intriguing, contradictory qualities of mathematics may initially puzzle those who are leery about the subject, but by exploring the history of its development, we can gain remarkable insights into mathematics' nature. With that in mind, we offer in *Math Makers* an overview of the history of mathematics, which we present through brief and exciting biographies of fifty of the most famous mathematicians, as well as clear investigations of some of their brilliant achievements.

As you consider the history of mathematics, you might ask questions such as:

Where did our current number system come from?
Whom do we credit for the beginnings of algebra and geometry?
Who was responsible for measuring the size of the earth—and how was it done with primitive tools?
Who invented the calculus?
What were the beginnings of calculators and computer programming?

In the biographies of these innovators, you will find the answers to not just these questions but also many more. Furthermore, the life stories of these men and women who invented and developed mathematics will both motivate you and inspire within you a greater appreciation for this most important subject.

Selecting which mathematicians to profile was no mean feat. We aimed for as broad a representation as possible, looking to feature specifically those who paved the path to our current technological age. This, of course, includes the all-too-often-neglected women who have contributed significantly to this process. Although each of these figures had markedly different life experiences, you will find a common characteristic among them: they were often considered unable to blend into the social fabric of the culture of their times. The brilliance and unusualness of these fifty mathematicians are revealed not only by the fruits of their mathematical wonder and labor but also by the very lifestyles they led.

Some of their lives were rather sad, such as that of French mathematician Évariste Galois, the developer of what is today known as Galois theory. In 1832, on the eve of a duel he believed himself sure to lose, the twenty-year-old Galois wrote down everything he knew about abstract algebra. Sadly, the duel eventually cost him his life. What he wrote that night became the foundation of Galois theory, which, as you will later see, connects two other theories in such a way as to make them both more understandable and simpler. One wonders what other gems Galois could have offered, were he given the chance.

But Galois was not the only mathematician whose contributions might have been lost entirely. In eighteenth-century European society, women were not allowed to participate in advanced academic studies. One of the famous mathematicians profiled here, Sophie Germain, was a child prodigy. In order to secure access to the world of academia, Germain wrote under the name of a former (male) student. After recognizing her genius and inquiring further, famous mathematicians of the day—such as Joseph Louis Lagrange and Carl Friedrich Gauss—discovered that she was a woman. Fortunately—and to our shared benefit—they accepted her as an equal. Germain then went on to provide significant advances in both mathematical studies and physics.

Another unusual, and rather melancholy, biography is that of the Indian mathematician Srinivasa Ramanujan. He grew up in very poor circumstances but was eventually accepted by famous British mathematicians. Yet he suffered poor health, which severely limited his life span. His biography was deemed worthy of a full-length feature film; 2014 saw the release of *The Man Who Knew Infinity*,[2] which was based off of the biography penned by Robert Kanigel.[3]

Perhaps one of the most unconventional lives detailed here was that of the Hungarian-American mathematician Paul Erdős, who essentially lived out of a suitcase. Erdős had no residence and lived with about five hundred mathematicians and universities for weeks at a time, and he published over 1,500 mathematical papers of high significance. Today, there still exists a pride among mathematicians who had the privilege of coauthoring a research article with him. The Erdős Number Project oversees the breakdown of who collaborated with this prolific mathematician and assigns Erdős numbers to them. Direct coauthors of his are designated as an "Erdős number 1"; coauthors of these mathematicians are then each considered an "Erdős number 2," and so on.

In the pages that follow, we survey not only modern mathematicians who advanced our shared knowledge but also those pioneering ancients who provided the foundation upon which the rest stood. For instance, Archimedes of Syracuse is mostly remembered as an ingenious inventor of mechanical devices; but he is also considered the greatest mathematician of classical antiquity. His mathematical achievements go well beyond the work of other ancient Greek mathematicians. Amazingly, he anticipated modern calculus when he used minute measurements to prove geometrical theorems. Such astounding accomplishments are pervasive in the biographies of these historical figures.

Beyond these awe-inspiring accomplishments, there are also many curiosities—some quite entertaining—that are part of the history of mathematics. For example, in 1637, in the margin of an algebra book, the famous French mathematician Pierre de Fermat wrote that no three positive integers a, b, and c satisfy the equation $a^n + b^n = c^n$ for any integer value of n greater than 2, but then also indicated that he did not have enough space in the margin to prove this conjecture. We know that when $n = 2$, this statement is true, as it is the well-known Pythagorean theorem. During the next 358 years, many famous mathematicians unsuccessfully attempted to prove Fermat's statement to be true, although no one ever found a counterexample. Hundreds of years later, a proof was finally provided by Andrew Wiles, in 1995; however, Wiles achieved this using methods certainly unknown to Fermat.

Another famous English mathematician, Christian Goldbach, made a conjecture in 1742 that still has not been proven true for *all* cases, but no one has yet found a case for which it doesn't hold true. His conjecture was written in a letter to the famous Swiss mathematician Leonhard Euler, and it is very simple—so much so that it could be easily understood by an elementary-school student. It states that every even integer greater than 2 can be expressed as the sum of two prime numbers. The attempts to prove this conjecture have led to many discoveries in the theory of numbers; but, to this day, the conjecture remains unproven for all cases.

As we guide you through this journey of the history of mathematics via the lives of those responsible for it, we explore also the work and developments for which they are famous. In some cases, we had to make judgments about what we would present as the highlights of a mathematician's achievements. This was particularly difficult with the biography of Leonhard Euler, who is known as the most prolific mathematician in history. As much as possible, we selected those works and achievements that are comfortably intelligible for the average person. This is consistent with our goal to make mathematics accessible, entertaining, and enjoyable, while at the same time appreciating the men and women who have discovered and presented the power and beauty of mathematics. After you become familiar with these remarkable individuals and their achievements, you will undoubtedly feel motivated to learn more about those who most particularly intrigued or inspired you. Their life stories encourage all of us to continue examining the world around us and how it is supported by this fascinating field of study. Furthermore, with a greater understanding of and respect for the most unique makers of our technological world, we also gain a deeper insight and ability to recognize the brilliance among outstanding people in our current society.

CHAPTER I

Thales of Miletus: Greek (ca. 624–546 BCE)

As we look back to the mathematicians of ancient times, we find that there is not much information regarding the details of their lives. What we do have is often a collection of contemporary commentaries written about them and perhaps some of their actual writings. We shall begin with one of the earliest of the outstanding major mathematicians, Thales of Miletus,

Figure 1.1. Thales of Miletus. (Illustration from Ernst Wallis et al., *Illustrerad verldshistoria utgifven*, vol. 1, *Thales* [Stockholm: Central-Tryckeriets Förlag, 1875–1879].)

who was born in 624 BCE in the ancient Greek city of Miletus (today Milet, Turkey). Although he had influence in the very early study of geometry, he is probably best known today for what we refer to as Thales's theorem, which simply says that if a triangle is inscribed in a circle with one side being the diameter of the circle, then the triangle is a right triangle. In addition to establishing this theorem, Thales led a very productive life not only as a mathematician but also as a philosopher and an astronomer, a combination that was common in his day.

The Greek society in which Thales was reared was less advanced than the societies of the ancient Egyptians and Babylonians, both of which cultures were leaders in mathematics and astronomy at the time. Despite this, it is believed that Thales was the Greeks' first true scientist. In his youth, Thales spent his time as a merchant, supporting his family's business.[1] His travels brought him to Egypt, which is where he most likely became enchanted with science and mathematics. He gradually reduced his thinking about spiritual influences on life and replaced it with scientific explanations. This change of interests significantly reduced his earnings but did not seem to stop him. Furthermore, Thales occasionally used scientific knowledge to his advantage in the business world. It is said that during a particular winter he realized that the coming season would have a bumper crop of olives, and, as a result, he secured all the olive presses in the region so that his potential competition was at a strong disadvantage. This is merely one example how, with a scientific understanding, he did earn quite a sum of money.

Let's look at some of the achievements in mathematics that are attributable to Thales. As we mentioned earlier, today he is best known for his accomplishments in geometry, since it is believed that he was the first to use deductive logic in establishing some geometric truths. In other words, he formalized the study of geometry from the typical practical aspects to the more formal deductive logic. One might say that Thales opened the doors for the study of geometry in ancient Greece, which peaked about three hundred years later. Thales died in the year 546 BCE, after having spent the last part of his life teaching at the Milesian school, which he founded.

Thales is largely remembered today for the theorem that bears his name. Although there are numerous ways to prove the theorem, we shall present one here that uses simple elementary geometry. In figure 1.2, we are

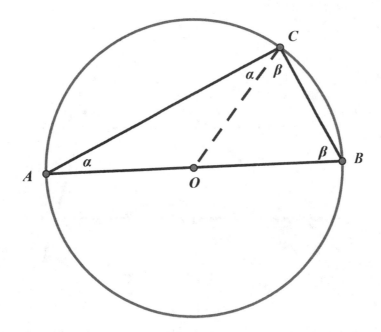

Figure 1.2.

given a triangle *ABC* inscribed in circle *O*, with side *AB* the diameter of the circle. Thales proved that angle *ACB* must be a right angle.

Since triangle *AOC* is an isosceles triangle, the base angles, that is, those marked with α are equal. Similarly, triangle *COB* is also an isosceles triangle, so that the two angles marked with β are also equal. Since the sum of the angles of a triangle is equal to 180°, we have the following: α + (α + β) + β = 180°. Then, 2α + 2β = 180°, or α + β = 90°, which is what we wanted to prove. Clearly, the converse is also true; namely, that the center of a circumcircle of a right triangle is on the hypotenuse of the right triangle.

Another theorem that is attributed to Thales is shown in figure 1.3, where parallel lines *AB* and *CD* are cut by two transversal lines *PCA* and *PDB*. Thales proved that the following proportions are true:

$$\frac{PC}{PA} = \frac{PD}{PB} = \frac{CD}{AB}.$$

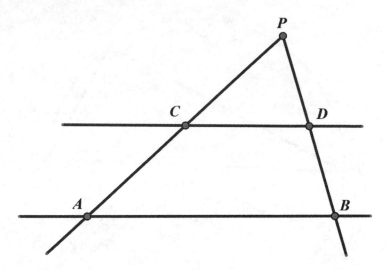

Figure 1.3.

These demonstrations give us a good insight into the new kind of thinking that Thales introduced to the world; in this sense, he was a trendsetter!

CHAPTER 2

Pythagoras:
Greek (575–500 BCE)

The one mathematician whom most people remember from their early school days is Pythagoras, whose name is attached to a theorem. As we embark on our exploration of the Pythagorean theorem, we are faced with some questions. Chief among them is, Why is the relationship that historically bears his name—the Pythagorean theorem—so important? There are many potential reasons: it is easy to remember; it can be easily visualized; it has fascinating applications in many fields of mathematics; and it is the basis for much of mathematics that has been studied over the past millennia. Yet, it may be best to begin at its roots—with the mathematician whom we credit as being the first to prove this theorem—and examine the man himself, his life, and his society.

When we hear the name Pythagoras, the first thing that pops into our minds is the Pythagoreans theorem.[1] When asked to recall mathematics instruction somewhat beyond arithmetic, it is common to remember that $a^2 + b^2 = c^2$. Those with a sharper memory may recall that this could be stated geometrically: the sum of the areas of the squares drawn on the legs of a right triangle is equal to the area of the square drawn on the hypotenuse. We can see this clearly in figure 2.1, where the area of the shaded square is equal to the sum of the areas of the two unshaded squares.

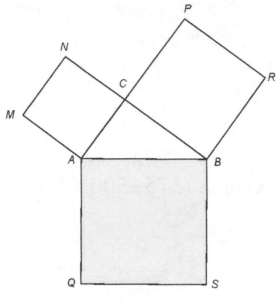

Figure 2.1.

There is probably no accurate picture of Pythagoras available today; however, the first biography of him was written about eight hundred years after his death. It was authored by Iamblichus, one of many Pythagoras enthusiasts, who tried to glorify him. Furthermore, although throughout history Pythagoras had been mentioned many other times by well-known writers, such as Plato, Aristotle, Eudoxus, Herodotus, Empedocles, and others, we still do not have very reliable information about him. Some of his contemporary followers actually believed that he was a demigod, a son of Apollo, a conviction they supported by noting that his mother was a very beautiful woman. Some reported that he even worked wonders.[2]

But just as he was called the greatest mathematician and philosopher of antiquity by some, he was not without critics who reviled him. The latter claimed that he was merely the founder and chief of a sect—the Pythagoreans; undermining the praise of him by authors, these critics argue that the many scientific results that came from this sect were written by its members and dedicated to its leader, thus, they were not the work of Pythagoras himself. The critics considered him a collector of facts without any deeper understanding of the related concepts; therefore, they believed that he did not truly contribute to a deep understanding of mathematics. Similar criticism also was aimed at such luminaries as Plato, Aristotle, and Euclid. This

Figure 2.2. Pythagoras depicted in
The School of Athens (fresco, Rafael, 1509–1511).

is seen throughout these historical recollections. We must remain mindful of these uncertainties when we consider the "facts" about Pythagoras' life and work.

Pythagoras was born in roughly 570 BCE on the island of Samos (located on the west coast of Asia Minor). His initial and perhaps most influential teacher was Pherecydes, who was primarily a theologist; Pherecydes taught religion, mysticism, and mathematics to Pythagoras. As a young man, Pythagoras traveled to Phoenicia, Egypt, and Mesopotamia, where he advanced his knowledge of mathematics and pursued a variety of other interests, such as philosophy, religion, and mysticism. Some biographers believe that, in his late teens, Pythagoras traveled first to Miletus, a town in Asia Minor near Samos, where he continued his studies in mathematics under the tutelage of the famous philosopher and mathematician Thales of Miletus. It is very likely that he also attended lectures from another Miletic philosopher, Anaximander, who further inspired Pythagoras in

geometry. When he returned to Samos at the age of thirty-eight, Polycrates
had come to power; this tyrant ruled Samos from 538 to 522 BCE. We are
not sure whether that is what prompted Pythagoras to leave Samos, since
soon thereafter, in about 530 BCE, he moved to Croton (today known as
Crotone, in southern Italy).

In Croton, Pythagoras founded a community—or society—whose
main interests were religion, mathematics, astronomy, and music (or acoustics). Members of this community became known as the Pythagoreans. The
Pythagoreans' goal was to explain the nature of the world, using numbers.
Specifically, they held a strong conviction that all aspects of nature and the
universe could be described and expressed by means of the natural numbers and the ratios of those numbers. This belief, however, suffered a set-
back when the society learned that the very emblem of their community—
the pentagram (a symmetric, five-cornered star)—contradicted their core
numerical principles.

One consequence of their overriding belief in the connectedness between the natural world and natural numbers would have been that, in particular, every two lines would have a common measure, that is, they would
be commensurable. Two magnitudes, a and b, are called commensurable if
there exists a magnitude m and whole numbers α and β such that $a = \alpha \cdot m$
and $b = \beta \cdot m$. But in the pentagon that encloses the pentagram, the sides
and the diagonals are *not* commensurable! In simpler terms, if we take the
length of one of the sides and divide it by the length of the diagonal, we
would not end up with a rational number, one that can be expressed as a
fraction. It is said that Hippasus of Metapontum, a student of Pythagoras,
discovered this fact and mentioned it to people outside of the community.
His actions were regarded as a violation of the society's pledge of secrecy,
so Hippasus was subsequently banned from the community. Some say that
he died in a shipwreck, which was then regarded as a punishment from the
gods for his sacrilege. Another version of the verbal reports holds that he
was killed by other members of the society. Clearly, this conviction held by
the Pythagoreans was one they took very seriously.

Beyond geometry and the relationships of lines to each other, the Pythagoreans held up for consideration many other aspects of the universe
and natural world. Acoustics was one of these. In an effort to discover its
connection to natural numbers, they studied vibrating strings. They found
that two strings sound harmonious if their lengths can be expressed as the
ratio of two small natural numbers, such as 1:2, 2:3, 3:4, 3:5, and so on.

Finding evidence such as this in many of their analyses, the Pythagoreans came to believe firmly that the entire universe must be ordered by such simple relations of natural numbers—hence, the seemingly severe reaction of the banning of Hippasus when he not only disproved their core belief but also shared that information with others.

Another core tenet of the society's philosophy was its belief that there is a strong connection between religion and mathematics. Pythagoreans believed that the sun, the moon, the planets, and the stars were of a divine nature; therefore, these celestial bodies could move only along circular paths. Furthermore, the followers of Pythagoras believed that the movements of these bodies created sounds of different frequencies, as a result of their different velocities, which in turn depended on each particular body's radius. These sounds were said to generate a harmonic scale, which they called the "harmony of the spheres." Yet, they believed that humans cannot actually hear this sound, as it surrounds us constantly, beginning from birth. Even the great German scientist Johannes Kepler (1571–1630) was sometimes characterized as a late Pythagorean, since he believed that the diameters of the orbits of the planets could be explained by inscribed and circumscribed Platonic solids (see fig. 2.3). Platonic solids are those solids whose surfaces consist of regular polygons of the same type (e.g., all equilateral triangles); there exist only five Platonic solids.[3] Kepler's idea regarding planetary orbits and Platonic solids was published in his work *Harmonices Mundi* ("The Harmony of the World") in 1619.

Part of what drew a large following to Pythagoras was that he was an eloquent speaker—in fact, four of his speeches, given to the public in Croton, are still remembered today.[4] In time, the Pythagoreans gained political influence in that region, even over the non-Greek population. But—as is frequent in politics—they faced resistance and animosity at times. For instance, later (in approximately 510 BCE), the Pythagoreans were involved in various political disputes, then were expelled from Croton. The society tried to move to other towns, such as Locri, Caulonia, and Tarent, but the locals did not allow them to settle. Finally, they found a new home in Metapontium. That is where Pythagoras eventually died of old age, in around 500 BCE.

Because there was no appropriately charismatic leader to succeed Pythagoras, the society split up into several small groups and tried to proceed with their tradition, while continuing to exert political influence in various towns in southern Italy. They were rather conservative and well connected to established influential families, which put them in conflict with their

Figure 2.3. Platonic solids. (Illustration from Johannes Kepler, *Mysterium Cosmographicum* ["The Cosmographic Mystery"] [Tübingen, 1597].)

common counterparts. As soon as their opponents gained the upper hand, bloody persecutions of the Pythagoreans began. Given this dire political situation, many Pythagoreans immigrated to Greece. This was—more or less—the end of the Pythagoreans in southern Italy. Very few individuals tried to continue the tradition and to advance the Pythagorean ideals. Two groups that persisted were the Acusmatics and the Mathematics, which in the ancient days meant "teacher" and later was used to indicate "that which was learned." The former group believed in acusma (i.e., what they had *heard* Pythagoras say), and did not give any further explanation. Their only justification was "He said it." This gave Pythagoras a level of importance, or popularity, in his day, which to some extent still persists. In contrast to the Acusmatics, the Mathematics tried to develop his ideas further and provide precise proofs for them.

One of the very few Pythagoreans who remained in Italy was Archytas of Tarentum (ca. 428–350 BCE). He was not only a mathematician and philosopher but also a very successful engineer, statesman, and military leader. He befriended Plato in about 388 BCE, which gave rise to the belief that Plato learned the Pythagorean philosophy from Archytas, and that that is why he discussed it in his works. Aristotle, who was first a student in Plato's academy but soon became a teacher there, wrote rather critically about the Pythagoreans. While Plato may have adopted many ideas from the Pythagoreans, such as the divine nature of planets and stars, in other cases he disagreed with them. Plato mentioned Pythagoras only once in his books, but not as a mathematician, despite his being in close contact with all of the mathematicians of his time and holding them in high regard.[5] It is probable that Plato did not consider Pythagoras a proper mathematician.

Similarly, Aristotle mentioned the Pythagoreans but said almost nothing about Pythagoras himself.[6]

In the fourth century BCE, the Greeks distinguished between "Pythagoreans" and "Pythagorists." The latter were extremists of the Pythagorean philosophy and consequently often the target of sarcasm because of their unusual, ascetic lifestyle. Still, among the Pythagoreans there were some members who were able to command respect from outsiders.

After the fourth century BCE, the Pythagorean philosophy disappeared until the first century BCE, when Pythagoras came into vogue in Rome. This "Neo-Pythagoreanism" remained alive for subsequent centuries. In the second century, Nicomachus of Gerasa wrote a book about the Pythagorean number theory, whose Latin translation by Boethius (ca. 500 BCE) was widely distributed. Today, Pythagorean ideas permeate our thinking in a variety of fields, and the Pythagorean theorem can be applied and proved in a wide variety of ways.

For example, suppose we begin with a square with its sides partitioned into segments of lengths a and b, as shown in figure 2.4, where we have the square divided into rectangles, triangles, and two smaller squares. We then move the four right triangles into the position of a congruent square, as shown in figure 2.5. We know that the acute angles of a right triangle have a sum of 90°; therefore, the figure in the center of this square (fig. 2.5) is also a square with sides of length c. The two large squares in figures 2.4 and 2.5 are congruent—each has sides of length $a + b$, and the sum of the areas of the four congruent right triangles in each of the two figures are equal. Therefore, the two smaller (unshaded) squares in figure 2.4—which have a combined area of $a^2 + b^2$—must have the same area as the unshaded square in figure 2.5, that is, c^2. Thus, we have $a^2 + b^2 = c^2$, and the Pythagorean theorem is proved!

For someone adept at elementary algebra, figure 2.5 nicely leads to the Pythagorean theorem in yet a different way. The area of the entire figure can be expressed in two ways:

1. you can find the area of the large square by squaring the length of a side ($a + b$) to get $(a+b)^2 = a^2 + 2ab + b^2$, or
2. you could represent the area of the large square as the sum of the areas of the four congruent right triangles, $4\left(\frac{1}{2}ab\right)$, plus the smaller inside square, c^2. This is:

$$4\left(\frac{1}{2}ab\right) + c^2 = 2ab + c^2 .$$

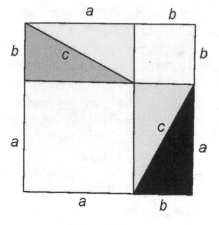

Figure 2.4.

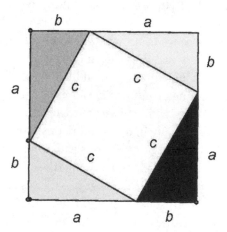

Figure 2.5.

Because you have two representations of the large square, you can simply equate them. Therefore, $a^2 + 2ab + b^2 = 2ab + c^2$. Then, by subtracting $2ab$ from both sides of the equation, you end up with the simple equivalent of $a^2 + b^2 = c^2$. This is the Pythagorean theorem as applied to the sides of any of the four congruent right triangles.

Today, there are more than 400 proofs of the Pythagorean theorem. In 1940, the American mathematician Elisha S. Loomis (1852–1940) published a book containing a collection of 370 proofs of the Pythagorean theorem done by many of the most famous mathematicians in history.[7] Loomis also notes that none of the proofs uses trigonometry. Students of mathematics know that all of trigonometry depends on the Pythagorean theorem; therefore, proving the theorem with trigonometry would be circular reasoning. Loomis's book also includes proofs provided by students and professors throughout the United States, as well as one presented by a United States president, James A. Garfield, which was published in 1876 in the *New England Journal of Education* under the title "Pons Asinorum."[8] Garfield's proof is a very interesting example of in how many ways this most popular theorem can be proved; therefore, we present it here.

In 1876, while still a member of the House of Representatives, the soon-to-be twentieth president of the United States, James A. Garfield, produced the following proof. Garfield was previously a professor of classics and, to this day, he has the distinction of being the only sitting member of the House of Representatives to have been elected president of the United States. Let's take a look at the proof he discovered.

In figure 2.6, $\triangle ABC \cong \triangle EAD$, and all three triangles in the diagram are right triangles.

Recall that the area of the trapezoid $DCBE$ is half the product of the altitude $(a + b)$ and the sum of the bases $(a + b)$, which we can write as $\frac{1}{2}(a+b)^2$. We can also obtain the area of the trapezoid $DCBE$ by finding the sum of the areas of each of the three right triangles:

$$\frac{1}{2}ab + \frac{1}{2}ab + \frac{1}{2}c^2 = 2\left(\frac{1}{2}ab\right) + \frac{1}{2}c^2 .$$

We can then equate the two expressions, since each represents the area of the entire trapezoid:

$$2\left(\frac{1}{2}ab\right) + \frac{1}{2}c^2 = \frac{1}{2}(a+b)^2 .$$

This can be simplified to $2ab + c^2 = (a+b)^2$, which can be written as $2ab + c^2 = a^2 + 2ab + b^2$ or, in more simplified form, $c^2 = a^2 + b^2$. This is the Pythagorean theorem as applied to right triangle ABC.

An astute reader may notice that Garfield's proof is somewhat similar to the one believed to be used by Pythagoras (see fig. 2.5). If we "complete" a square from the given trapezoid in figure 2.6, we get a configuration similar to that in figure 2.5. This "completed square" is shown in figure 2.7.

And so we now have a little bit of history about the man we claim is responsible for what many people consider to be perhaps the most famous theorem in mathematics. The Pythagorean theorem is the one most people remember when they think back to their school years in mathematics.

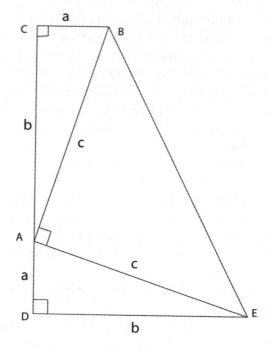

Figure 2.6.

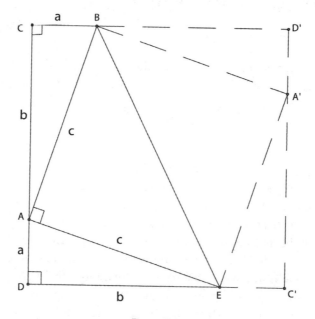

Figure 2.7.

CHAPTER 3

〜

Eudoxus of Cnidus: Greek (390-337 BCE)

For many years it was believed that both Isaac Newton (1642–1726) and Gottfried Wilhelm Leibniz (1646–1716) invented calculus. Newton's work was called fluxions, and Leibniz's notation is what is used today in the study of differential and integral calculus. However, research in modern times has shown that the actual "inventor" of what we today call calculus was, in fact, Eudoxus of Cnidus, who was born in Cnidus, Asia Minor, and around 390 BCE. His work, which is today considered the forerunner of calculus was called *Method of Exhaustion*. Eudoxus is often seen as the greatest of the classical Greek mathematicians, with the possible exception of Archimedes. Unfortunately, all of his written work seems to have been lost over the years; however, his work is cited by many mathematicians who followed him, including Euclid.

Most of what we know about Eudoxus's life comes from the third-century historian Diogenes Laertius, who wrote a compilation of biographical snippets—along with some gossip—which included Eudoxus among the many other famous philosophers and mathematicians.[1] From Laertius, we know that at age twenty-three, while in Athens, Greece, Eudoxus was to have attended lectures at Plato's Academy. Soon thereafter, he left for Egypt, where he spent sixteen months studying with priests and making astronomical observations from an observatory. In order to support himself, he did some teaching and returned to Asia Minor; later, he returned to Athens, where he worked at the Platonic Academy as a teacher. Eventually,

Figure 3.1. Eudoxus of Cnidus.

he returned to Cnidus, where he became a legislator and continued doing research. He died in about 337 BCE.

In Book V of Euclid's *Elements*, much of the discussion of proportionality seems to be credited to Eudoxus; however, it is not known to what extent subsequent mathematicians' work was included in the discussion. During this time, Greek mathematicians measured objects via proportionality, that is, the ratio of two similar items was compared to others in the same way, thereby forming a proportion. This was unlike our modern-day methods of measuring quantities either numerically or through various equations. Eudoxus is credited with giving meaning to the equality of two ratios, or a proportion. Euclid's Book 5, definition 5, of *Elements*, which is largely credited to Eudoxus, reads as follows:

> Magnitudes are in the same ratio, that is, the first to the second and the third to the fourth when, if any equal multiples are taken of the first and the third, and any equal multiples of the second and fourth, the latter equal multiples exceed, or are equal to, or are less than the latter equal multiples, respectively, taken in corresponding order.

This could be more easily explained symbolically in the following way: Consider the four quantities a, b, c, and d. Now consider the ratios $\frac{a}{b}$ and $\frac{c}{d}$. Let us now consider them equal, so that $\frac{a}{b} = \frac{c}{d}$. Now we consider two arbitrary numbers, say, p and q, and form multiples of the first and third numbers to get pa and pc. Similarly, we form multiples of the second and fourth numbers to get qb and qd. If $pa > qb$, then it must follow that $pc > qd$. On the other hand, if $pa = qb$, then it must follow that $pc = qd$. Furthermore, if $pa < qb$, then it must follow that $pc < qd$. Bear in mind that these definitions referred to comparing similar quantities, not necessarily similar units of measure. Most important, Eudoxus's definition does not require a, b, c, and d to be rational numbers; his definition of equality of two ratios also works for irrational numbers.

Furthermore, Pythagoreans had discovered that there exist numbers that cannot be expressed as a ratio, $\frac{p}{q}$, where p and q are integers. Their method of comparing two lengths a and b was to find a length u so that $a = p{\cdot}u$ and $b = q{\cdot}u$ for whole numbers p and q. It had been thought that for any two lengths a and b there always exists some sufficiently small unit u that could fit evenly into one of these lengths as well as the other. However, the Pythagoreans were upset when they found out that such a common unit of measure does not always exist; not all lengths can be compared or measured in this way. For example, the length of the hypotenuse of an isosceles right triangle with legs of length 1 is incommensurable with its legs, meaning that there exists no unit of measure u that would fit evenly into both the hypotenuse and the leg. By the Pythagorean theorem, the length of the hypotenuse of this triangle is $\sqrt{2}$. Saying that this number is incommensurable with 1 means that it is impossible to have $\sqrt{2} = m \cdot u$ and $1 = n \cdot u$ with the same number u in both equations and with both m and n whole numbers; that is, $\sqrt{2}$ and 1 cannot be expressed as multiples of a common unit of measure. In other words, $\sqrt{2}$ cannot be written as a ratio of whole numbers, $\frac{m}{n}$; therefore, it is not a rational number—it is irrational.

As mentioned above, Eudoxus's method of comparing ratios allows us to compare or measure irrational numbers; in this sense, he was the first who made irrational numbers measurable. In fact, the German mathematician Richard Dedekind (1831–1916) emphasized in his writings that he was inspired by the ideas of Eudoxus when he developed the notion known as a Dedekind cut, which is now a standard definition of the real numbers. The idea of a Dedekind cut is that an irrational number divides the rational numbers into two classes, or sets, with all the numbers of one (greater) class being strictly greater than all the numbers of the other (lesser) class. For example, $\sqrt{2}$ divides into the lesser class all the negative numbers and the numbers the squares of which are less than 2; divided into the greater class, then, are the positive numbers the squares of which are greater than 2.

Beyond rendering irrational numbers measureable, as indicated earlier, Eudoxus is also credited with having developed the method of exhaustion. Exhaustion is a process for finding the area of the shape by inscribing within it a series of polygons—with ever-increasing number of sides—whose areas eventually converge to the area of the original figure. When constructed directly, the difference in area between the nth polygon and the original shape being measured will become smaller as n becomes larger. As this difference becomes arbitrarily small, the area of the original shape is eventually "exhausted" by the lower-bound areas, successively established by the sequence members. Again, the method of exhaustion preceded integral calculus. It did not use limits, nor did it use infinitesimal quantity. It was merely a logical procedure based on the idea that a given quantity can be made smaller than another given quantity by continuously halving it a finite number of times. One example of this would be to show that the area of a circle is proportional to the square of its radius.

Although the true study of calculus originated through the writings of Isaac Newton and Gottfried Wilhelm Leibniz, we must credit Eudoxus for having established with the method of exhaustion a forerunner of today's calculus. Understanding Eudoxus's contributions to the mathematics we know and use today grants us a true appreciation not only for his individual forethought and inventiveness but also for how all learning is cumulative. The vast and astounding achievements we take for granted today would be impossible or nonexistent without the work and brilliance of those who preceded us and provided a foundation upon which we and our more recent forebears could build.

~

CHAPTER 4

Euclid:
Greek (ca. 300 BCE)

No collection of extraordinary mathematicians would be complete without including Euclid, who was often known as Euclid of Alexandria (a city in Hellenistic Egypt). Although there is hardly any evidence about his life available, it is believed that he lived around 347 BCE. Centuries later, Euclid was popularized by the Greek philosopher Proclus Lycaeus (ca. 450 CE).[1] What little is known about his life is that he probably received his mathematical training in Athens, from Plato's pupils, since most of the geometers seem to have gravitated there. Even Euclid's time in Alexandria is not clearly defined, since the Greek mathematician Apollonius, who flourished around 200 BCE, makes reference to Euclid in the introduction to his book *Conics*. Therefore, we deduce that Euclid must have lived prior to that time, and the reference provides further evidence of the importance of his book *Elements*. Beyond this book on geometry, Euclid was also the author of a book on optics, which he approached from a geometric standpoint.

Although Euclid is best known for the *Elements*, there is no original copy of this work. There is also a persistent belief that the *Elements*—which consists of thirteen books—was merely a work that Euclid wrote as a collection of material that had been previously developed by many mathematicians. In any case, the *Elements* is one of the most important works in mathematics, and although it is largely a study of geometry, it also includes a fair amount of number theory. What makes the *Elements* so remarkable is to a lesser extent previously unknown mathematical results contained in this work, but much

Euclides Megarensis Philosophus clarus habetur olympiade 90. temporibus Darij Nothi Perfarum Regis quinti.

Figure 4.1. Euclid.

more the organization of the material. Euclid begins with definitions and five postulates (axioms), followed by theorems and their proofs. All theorems are derived from the five axioms stated at the beginning. Throughout the book he kept a very high level of rigor, dramatically raising the standard for any mathematical work to be written in the future. The clarity with which the theorems are stated and proved is unprecedented. The Elements basically defined the style of modern mathematical literature (see fig. 4.2).

The traditional geometry course that today is offered in most American high schools is based on the work of Euclid. It is, therefore, called Euclidean geometry, which refers to geometry on a plane (as opposed to, for example, geometry on the surface of a sphere). Perhaps the most significant principle that commands the geometry throughout the rest of *Elements* is Euclid's fifth postulate. It reads as follows:

If a line segment intersects two straight lines, forming two interior angles on the same side of that given line, such that sum of their

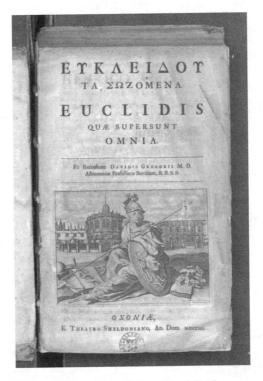

Figure 4.2. The 1704 edition of Euclid's *Elements*.

measures is less than two right angles, then the two lines, if extended indefinitely, will meet on that same side of the given line, where those two angles have a sum less than two right angles.

This was vastly simplified in 1846 by the Scottish mathematician John Playfair (1748–1819), who stated an equivalent postulate, known today as Playfair's axiom. It reads as follows:

In a plane, given a line and a point not on it, at most, one line parallel to the given line can be drawn through the point.

It is this axiom that today guides the basics in Euclidean geometry.

The high-school study of geometry in the United States is rather unique in the world today, in that an entire school year is devoted to the logical development of geometry. This essentially began with the Scottish mathematician Robert Simson's (1687–1768) classic geometry book titled *The*

Elements of Euclid. This book, published in 1756, offered the first English version of this classic work by Euclid; furthermore, it was also the basis for the study of geometry in England. In figure 4.3, we show the seventh edition from 1787.

Through this we can appreciate the notion that Euclid's influence transcended the study of geometry. Yet the study of geometry took its own path through Simson's English version, which was then adopted and modified by the French mathematician Adrien-Marie Legendre (1752–1833). In 1794, Legendre wrote a textbook titled *Eléments de géométrie*, which in turn became the model for the American high-school geometry courses as we know them today. It came to prominence in a rather circuitous route: first from Euclid, then to Simson, then to Legendre. Then, Legendre's book was translated from French by David Brewster in 1828 and titled *Elements of Geometry and Trigonometry*; from there it was adapted in 1862 by the American mathematician Charles Davies (1798–1876) as a school course,

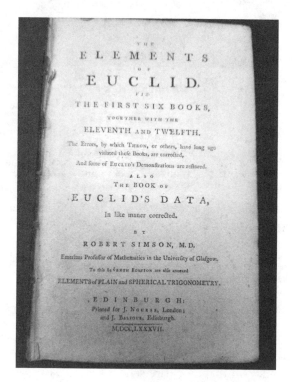

Figure 4.3. The seventh edition of *The Elements of Euclid*,
by Robert Simson, published in 1787.

although, in the early days this was also a college-level course. Simson was so popular as a geometer that there were even theorems named for him about which he knew nothing. For example, the famous geometry theorem carrying his name—the Simson line—was first developed by the Scottish mathematician William Wallace in 1799, well after Simson's death. It states that from any point on the circumscribed circle of a triangle, the feet of the perpendiculars drawn to each of the three sides are collinear. This can be seen in figure 4.4.

The *Elements* had great influence beyond the realm of mathematics, reaching across the ages to influence American history as well. For example, in his 1860 autobiography, President Abraham Lincoln stated about himself (in the third person), "After he was twenty-three and had separated from his father, he studied English grammar—imperfectly, of course, but so as to speak and write as well as he now does. He studied and nearly mastered the six books of Euclid, since he was a member of Congress."[2] Although the first six books have to do largely with geometry, they provided for Lincoln an ability to improve his mental faculties, especially his powers of logic and language. He even referred to Euclid in the famous fourth debate he had in 1858 with Senator Stephen A. Douglas (1813–1861) in Charleston, South Carolina. Referencing Euclid, he said:

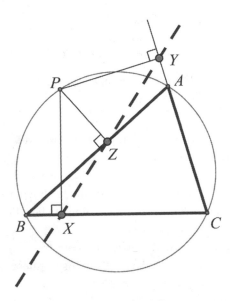

Figure 4.4. Simson's line.

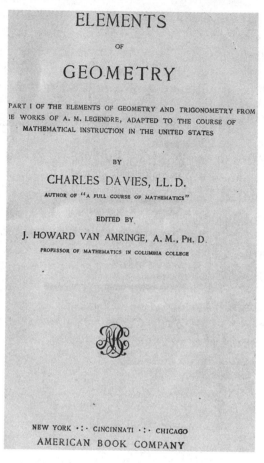

Figure 4.5. The first American geometry course book.

If you have ever studied geometry, you remember that by a course of reasoning, Euclid proves that all the angles of a triangle are equal to two right angles. Euclid has shown you how to work it out. Now, if you undertake to disprove that proposition, and to show that it is erroneous, would you prove it to be false by calling Euclid a liar?[3]

It was also known at the time that when Lincoln traveled by horseback, he always carried a copy of Euclid's *Elements* in his saddlebags. Although Lincoln had no formal education, we can see that his devotion to learning was truly remarkable, and the influence of Euclid was of particular note.

Although we have very little information about Euclid's biography, we can see the legacy that he initiated through his famous book *Elements*. To the present day, not only does it provide the basis for our high-school geometry studies, but it also has played a notable role in our logical thinking on the national level, as was exhibited by Abraham Lincoln's application of Euclid's reasoning. Through Euclid's influence, we see how the reach of these early mathematicians has spread beyond mathematics itself to encompass science, logic, philosophy, education, and more.

Chapter 5

Archimedes:
Greek (ca. 287–ca. 212 BCE)

There is no Nobel Prize for mathematics, but there are two awards in mathematics with a comparably high prestige, at least within the community. One is the Abel Prize, established in 2002 by the Norwegian government; the other is the Fields Medal, which was first awarded in 1936. Unlike the Nobel Prizes and the Abel Prize, which are awarded annually, the Fields Medals are awarded only every four years—and there is an age limit for its recipients. They must be under forty years of age. Although it might be strange to impose an age limit on such a prestigious award, there is a reason for doing so: the award is also intended to encourage future research. Officially known as the international medal for outstanding discoveries in mathematics, the colloquial name "Fields Medal" is in honor of the Canadian mathematician John Charles Fields (1863–1932). He began developing the award in the late 1920s and even chose the design of the medals. Unfortunately, he died from a stroke two years before the first medals were awarded. In his personal will, he left a $47,000 grant to establish a fund for the award. The Fields Medal is made of gold, and its front side shows the head of Archimedes[1] (ca. 287–ca. 212 BCE) and the inscription "*Transire suum pectus mundoque potiri*" (see fig. 5.1). This is a quotation attributed to Archimedes; it can be translated as "Rise above oneself and grasp the world."

Although this quote is emblazoned on the Fields Medal that bears his likeness, Archimedes is probably most famous for having proclaimed "*Eureka! Eureka!*" This particular phrase translates to "I've found it! I've found

Figure 5.1. The Fields Medal. (Image from Stefan Zachow of the International Mathematical Union, retouched by King of Hearts.)

it!" This is what he exclaimed after having stepped into a bath and suddenly noticing that the amount by which the water level rose was a measure of the volume of the part of his body he had submerged (i.e., displacement). The background story adjoining this anecdote is that the local tyrant, Hiero II of Syracuse (ca. 308–215 BCE) contracted Archimedes to find a method by which the purity of a golden crown could be assessed without destroying it. Hiero's request stemmed from his suspicion that his goldsmiths had replaced with silver some of the gold he had given them for the creation of the crown. Archimedes was able to solve the problem because gold weighs more than silver. Therefore, a crown mixed with silver would have to be bulkier than a purely golden crown of the same weight. Consequently, the adulterated crown would also displace more water. Although this story is compelling, the oldest source for it is a book on architecture by the Roman writer Vitruvius,[2] which appeared approximately two hundred years after the alleged episode; it is very likely that the story has been substantially modified and embellished, even though there may be some truth in it. Further undermining this tale's veracity, Galileo Galilei (1564–1642) pointed out that Archimedes could have achieved a much more accurate measurement by using a different method that relied on his own law of buoyancy, which is now known as Archimedes's principle. Archimedes is also remembered as an ingenious inventor of mechanical devices, such as Archimedes's screw for lifting water (see fig. 5.2) and various "super weapons" of the ancient world.

Perhaps less well known is the fact that Archimedes is generally considered the greatest mathematician of classical antiquity, which is why his profile decorates the Fields Medal. His mathematical achievements go well beyond the work of other ancient Greek mathematicians; in particular, he applied and perfected Eudoxus' method of exhaustion to prove geometrical theorems. He developed the proofs that anticipated those of modern calculus. Unfortunately, almost nothing is known about Archimedes's life, except for a few anecdotes and some biographical information he mentioned in his writings.

Archimedes was born in the city of Syracuse on the island of Sicily in roughly 287 BCE. His father was an astronomer named Phidias, of whom nothing else is known. It is believed that Archimedes studied in the intellectual center of the ancient world: Alexandria, Egypt. Supporting this belief is the fact that two of Archimedes's works have introductions addressed to Eratosthenes (ca. 276–ca. 195 BCE), who was in charge of the legendary Great Library of Alexandria, a place where the greatest scholars of the ancient world would meet. Eratosthenes is famous for calculating the circumference of the earth by measuring the sun's angle of elevation at noon both in Alexandria and in a city a known north–south distance away from Alexandria.

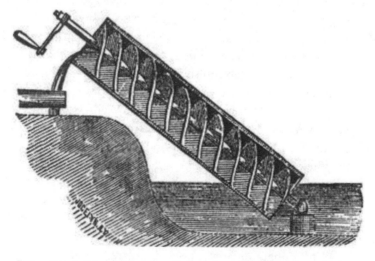

Figure 5.2. Archimedes's screw. (Image from *Chambers's Encyclopedia*
(Philadelphia: J. B. Lippincott, 1875.)

Figure 5.3. Engraving from the book *Les vrais pourtraits et vies des hommes illustres grecz, latins et payens* (1586).

Considering the extremely primitive measuring tools Eratosthenes had at his disposal, he obtained a remarkably accurate result (see chap. 6).

The Greek mathematician Conon of Samos (280–220 BCE) was another contemporary of Archimedes, and he, too, was mentioned in Archimedes's writings. Unfortunately, only fragments of transcriptions of Archimedes writings have survived, and at least seven of his treatises are completely lost. (Other authors referred to them, which is why we know that they must have existed.) However, the few copies of his treatises that survived through the Middle Ages were highly influential for scientists and mathematicians of the Renaissance, notably for Galileo Galilei (1564–1642), Johannes Kepler (1571–1630), René Descartes (1596–1650), and Pierre de Fermat (1607–1665).

While Archimedes often uses heuristic arguments—which employed logical problem-solving methods—to find the solution to a mathematical problem, he then also provides a rigorous proof for his result. In his treatise "The Method of Mechanical Theorems," he emphasizes that heuristic reasoning, although often very useful to obtain an "educated guess" for the solution, cannot replace a mathematical proof. In antiquity, Archimedes was famous for his inventions, but his mathematical writings were not so well known. Consequently, it was more than seven centuries until his works were first compiled into a comprehensive text now known as the Archimedes codex. The codex was compiled around 530 CE by Isidore of Miletus, an architect of the Hagia Sophia patriarchal church in the Byzantine Greek capital city of Constantinople (now Istanbul). The discovery of Archimedes's treatise "The Method of Mechanical Theorems" is a real-life *Indiana Jones* story, one well worth a short digression: A copy of the Archimedes codex was made around 950 CE by an anonymous scribe, but in the thirteenth century, the parchment leaves of this copy were reused for a Christian religious text. Parchment was very expensive and not readily available, so it was often "recycled" by scraping off the previous writing and then washing the pages. Mathematical texts that could only be understood by a handful of scholars were quite literally considered not worth the "paper" on which they were written. The original text on the pages of a manuscript that underwent this recycling procedure are called a palimpsest, derived from an ancient Greek compound word meaning "scraped clean to be used again." The cleaned leaves of the Archimedes codex were folded in half, so that each sheet became two pages of the liturgical book (see fig. 5.4).

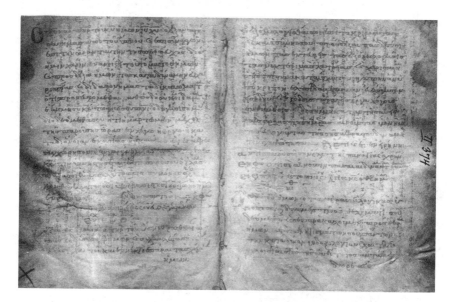

Figure 5.4. The Archimedes Palimpsest (Archimedes's text is the fainter one running from left to right).

Fortunately, the erasure was incomplete. In 1846, while studying ancient biblical texts in a Greek Orthodox library in Constantinople, the German biblical scholar Constantin von Tischendorf (1815–1874) discovered a faint mathematical text covered up by the religious writing in an old prayer book. He excised a sample page and took it with him, but he could not determine the value or meaning of the mathematical text obscured by the prayers. After Tischendorf's death, the University of Cambridge bought a collection of manuscript pages from his estate, including the palimpsest. At Cambridge, the unidentified sheet received a number and was filed, which could have been the end of the story. However, in 1899, a Greek scholar produced a catalog of the books in the Constantinople library, and he also discovered the faint mathematical text in the prayer book. He transcribed several lines of it, which were called to the attention of the world's leading expert on Archimedes, who realized that the text was indeed from a treatise of Archimedes. When he visited the library in 1906, he was permitted to take photographs of the palimpsest, from which he then produced transcriptions. The palimpsest included works by Archimedes that were thought to have been lost. This sensational discovery made headlines in the newspapers throughout the world. However, during the Greco-Turkish War (1919–1922), the palimpsest was stolen, and later it was sold to a businessperson who stored it in a cellar, where it remained for decades and was damaged by water and mold. In 1971, a single-sheet of the palimpsest, kept at Cambridge, was identified as belonging to the Archimedes palimpsest from the Constantinople library. In 1998, the book reappeared in the public eye at a Christie's auction, where it was sold to an anonymous buyer for $2 million. The severely damaged book was then taken to the Walters Art Museum in Baltimore for conservation, which was an extremely challenging task that took several years to complete.[3] Highly sophisticated imaging techniques were used to create a digital copy of the palimpsest before it was returned to its new owner.[4] The palimpsest contains the only existing copy of Archimedes's treatise "The Method of Mechanical Theorems," which was an extremely important discovery, since it provided new insights into how Archimedes obtained his results.

In this work, Archimedes shows how the area or volume of a figure can be determined by dissecting it into an infinite number of infinitely small parts (infinitesimals), thereby anticipating the modern concept of the integral. However, since he did not view the use of infinitesimals to be rigorous mathematics, Archimedes also provided proofs based on already-established methods. The proofs he gave in his treatise relied on the method of exhaustion and the *reductio ad absurdum* ("reduction to absurdity"), two techniques he brought to perfection. The *reductio ad absurdum* is a mode of reasoning in which one attempts to disprove a statement by showing that it inevitably leads to an "absurd" conclusion, for instance, a mathematical contradiction such as 1 = 0. This type of argument can be traced back to classical Greek philosophy, notably to Aristotle, and it was also applied by Euclid to prove mathematical theorems. While the *reductio ad absurdum* is not limited to mathematical reasoning and is used in philosophy, the method of exhaustion is of a purely mathematical nature. It modern terms, it consists of finding the area (or volume) of a shape by inscribing inside it a sequence of polygons (or polyhedra) with increased numbers of sides, whose areas (or volumes) converge to that of the given shape. To reveal these two methods in more detail, we shall sketch Archimedes's proof of a result he published in his treatise "Measurement of a Circle." Only a fragment of this work has survived; it consists of three propositions, the first of which states that the area of any circle is equal to the area of a right triangle in which one of the legs of the right angle is equal to the radius of the circle, and the other leg is equal to its circumference (see fig. 5.5). That is,

$$Area = \frac{1}{2}r\left(2\pi r\right) = \pi r^2 .$$

An essential element of Archimedes's proof of the formula for the area of a circle is the approximation of the circle by regular polygons inscribed in the circle. For example, figure 5.6 shows a square and an octagon inscribed in a circle.

![circle with radius r and circumference c, and a right triangle with legs r and c]

Figure 5.5.

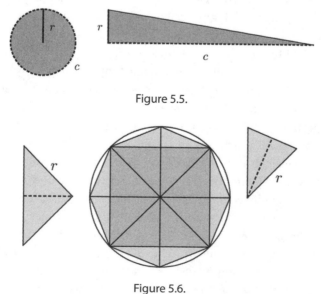

Figure 5.6.

The octagon can be obtained from the square by erecting isosceles triangles on the sides of the square, with the square's vertices touching the circle. Repeating this procedure with the octagon would produce a 16-gon. With each doubling of the number of sides, the area of the inscribed polygon increases, but it will always be less than the area of the circumscribed circle. However, we can approximate the area of the circle by the area of an inscribed n-gon with arbitrary precision, if we only take n large enough. Similarly, we could use circumscribed n-gons whose sides are tangent to the circle. An n-gon can be decomposed into n isosceles triangles (see, e.g., fig. 5.6); its area is, therefore, $Area_n = n \cdot \frac{base \cdot height}{2} = \frac{1}{2} c_n h_n$, where c_n is the perimeter of the n-gon and h_n is its apothem (the length of the segment from the center of the n-gon to the midpoint of one of its sides). Denoting the area of the circle and the right triangle shown in figure 5.5 by $Area_{Circle}$ and $Area_{Triangle}$, respectively, we want to show that $Area_{Circle} = Area_{Triangle}$. To this end, Archimedes used a *double* reductio ad absurdum: First, we assume that $Area_{Circle} > Area_{Triangle}$. If we take n sufficiently large, then the area of the inscribed n-gon will lie between the area of the circle and the area of the triangle, that is, $Area_{Circle} > Area_n > Area_{Triangle}$ (recall that we can make the approximation of the circle by a polygon as close as we wish, but $Area_n$ will always be smaller than $Area_{Circle}$). Since the legs of the right triangle have length r (radius of the circle) and c (circumference of the circle), we have $Area_{Triangle} = \frac{1}{2} rc$. On the other hand, the perimeter of the inscribed n-gon must be smaller than the circumference of the circle, and its apothem must be smaller than the radius of the circle, implying that $Area_n = \frac{1}{2} c_n h_n < \frac{1}{2} rc = Area_{Triangle}$, which is a contradiction to $Area_n > Area_{Triangle}$. Consequently, the assumption that $Area_{Circle} > Area_{Triangle}$ must have been wrong. If we now assume that $Area_{Circle} < Area_{Triangle}$, we may construct an n-gon circumscribed about the circle, such that $Area_{Circle} < Area_n < Area_{Triangle}$. Since we have $c_n > c$ and $r_n > r$ for any n-gon circumscribed about the circle, we obtain $Area_n = \frac{1}{2} c_n h_n > \frac{1}{2} rc = Area_{Triangle}$, in contradiction to $Area_n < Area_{Triangle}$. This implies that the assumption $Area_{Circle} < Area_{Triangle}$ must have been wrong as well. We thus have shown that neither $Area_{Circle} > Area_{Triangle}$ nor $Area_{Circle} < Area_{Triangle}$ can be true, from which we can conclude that $Area_{Circle} = Area_{Triangle}$.

Using the technique of inscribing and circumscribing regular polygons in and about a circle, Archimedes was also able to determine the value of π with remarkable accuracy. The number π is defined as the ratio between the circumference and the diameter of a circle. For an inscribed or circum-scribed n-gon, the ratio

$$\frac{perimeter}{2 \cdot apothem} = \frac{c_n}{2h_n}$$

will approach π as n gets larger and larger. Archimedes developed a numerical procedure to calculate the perimeter and apothem of inscribed and circum-scribed n-gons and carried out the calculation for $n = 12, 24, 48,$ and 96, thereby obtaining a lower and an upper limit for the value of π:

$$3\frac{10}{71} < \pi < 3\frac{1}{7} = \frac{22}{7} = 3.142 \ldots$$

(where $\pi = 3.14159265 \ldots$). The approximation of π by the fraction $\frac{22}{7}$ be-came very popular in antiquity and was commonly used in calculations un-til the Middle Ages. Today, this approximation is also quite popular among students in younger grades.

In "On the Sphere and Cylinder," Archimedes shows that the surface area of a sphere is four times that of a great circle (a circle on the surface of a sphere with its center at the center of the sphere). In other words, *Area* $= 4\pi r^2$, where r is the radius of the sphere. The volume contained in the sphere is two-thirds the volume of a circumscribed cylinder; or symboli-cally, $Volume_{Sphere} = \frac{4\pi r^3}{3}$, $Volume_{Cylinder} = 2\pi r^3$). Archimedes was very proud of this result and consequently left instructions for his tomb to be marked with a sphere inscribed in a cylinder. In fact, the Roman philosopher Cicero (106–43 BCE) visited Archimedes's tomb when he was in Sicily in 75 BCE, 137 years after Archimedes' death. After some searching, he found the tomb "enclosed all around and covered with brambles and thickets," and he wrote:

> I noticed a small column arising a little above the bushes, on which there was a figure of a sphere and a cylinder. . . .

Unfortunately, the location of Archimedes's tomb is not known. In an-other nod to this brilliant mathematician, the reverse of the Fields Medal displays a sphere inscribed in a cylinder (see fig. 5.7).

Archimedes's proof of the relationship between the volume of a sphere and a circumscribed cylinder is a masterpiece of mathematics. However, rather than provide a detailed version of his proof, we provide a less rigor-ous version, which will be more intuitively comprehensible. We will need the formula for the volume of a cone; therefore, we will first provide a heu-ristic argument for how to compute the volume of a cone. Consider a regu-lar square pyramid whose height is half the length of a side of its base. We can combine six of such pyramids to form a cube, as shown in figure 5.8.

Figure 5.7. The Fields Medal (reverse). (Image from Stefan Zachow of the International Mathematical Union, retouched by King of Hearts.)

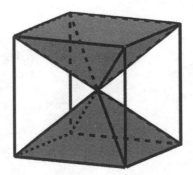

Figure 5.8.

Hence, the volume of one pyramid must be one sixth of the volume of the cube. If a is the length of a side of the cube, we obtain $Volume_{Pyramid} = \frac{1}{6} Volume_{Cube} = \frac{1}{6}a^3 = \frac{1}{3}a^2\left(\frac{a}{2}\right) = \frac{1}{3}Bh$, where B is the area of the base of the pyramid and h is the length of its height. We will now argue that the formula $volume = \frac{1}{3}(base\ area)(height)$ is also valid for arbitrary pyramids as well as for cones. Toward this end, we employ a theorem known as Cavalieri's principle. It is named after the Italian mathematician Bonaventura Cavalieri (1598–1647), but it is actually just a modern implementation of Archimedes's method of exhaustion. If we consider two regions in space included between two parallel planes, then Cavalieri's principle states that if every plane parallel to these two planes intersects both regions in cross sections of equal area, then the two regions have equal volumes. Cavalieri's principle is very intuitive and can be nicely illustrated with a stack of coins, as shown in figure 5.9. The volume of the stack does not change if we misalign the coins.

In fact, we may also melt a coin and make a triangular coin out of it, or any other shape; as long as the areas of the cross sections of a region stay the same, its volume is preserved. This implies that if we have two pyramids with the same base area, B and the same height, H, they must have the same volume—no matter whether they are oblique or irregular. The same is true for a cone (a cone can be thought of as the limit of a pyramid whose base is a regular n-gon, when n approaches infinity). For a cone as well as for any pyramid, the base area and the height are the only quantities relevant for calculating the volume, which is always equal to $\frac{1}{3}(base\ area)(height)$. We may now consider, as Archimedes did, a sphere inscribed in a cylinder. Archimedes noticed that the volume of the cylinder minus the volume of the sphere is exactly equal to the volume of a double cone, which is shown symbolically in figure 5.10.

Figure 5.9.

Figure 5.10.

To see that this is true, we just have to convince ourselves that at any height, the *sum* of the areas of the cross sections of the sphere and the double-cone equals the area of the cross section of the cylinder. In figure 5.11, we show the vertical projection of a sphere of radius r inscribed in a cylinder, together with the double cone. The cross section of the sphere at height h is a circle with radius $AD = \sqrt{r^2 - h^2}$. The cross section of the double cone at height h is a circle of radius $BD = h$. Since the area of a circle of radius R is equal to πR^2, the sum of the cross-sectional areas of sphere and double cone at height h is $\pi(r^2 - h^2) + \pi h^2 = \pi r^2$, which is exactly the cross-sectional area of the cylinder. Thus, by Cavalieri's principle, the volume of the cylinder is exactly equal to the sum of the volumes of the sphere and the double cone. Since the volume of the double cone is $2 \cdot \frac{1}{3}(base\ area) \cdot (height) = 2 \cdot \frac{1}{3}\pi r^2 \cdot r = \frac{2}{3}\pi r^3$ and the volume of the cylinder is $(base\ area) \cdot (height) = \pi r^2 \cdot 2r = 2\pi r^3$, the volume of the sphere must therefore, be $2\pi r^3 - \frac{2}{3}\pi r^3 = \frac{4}{3}\pi r^3$, which is two-thirds of the volume of the cylinder, as Archimedes has shown.

In his lifetime, Archimedes was much more famous for his mechanical inventions than for his outstanding and far-reaching work in mathematics, yet he was convinced that pure mathematics was the only worthy pursuit. His fascination with geometry in particular is beautifully described by the Roman writer Plutarch (46–120 CE):

Oftentimes, Archimedes' servants got him against his will to the baths,

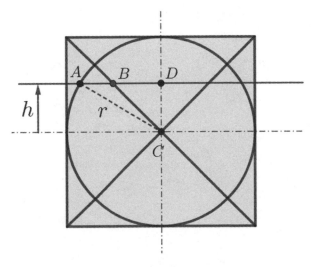

Figure 5.11.

to wash and anoint him, and yet being there, he would ever be drawing out the geometrical figures, even in the very embers of the chimney. And while they were anointing of him with oils and sweet savors, with his fingers he drew lines upon his naked body, so far was he taken from himself, and brought into ecstasy or trance, with the delight he had in the study of geometry.[5]

Archimedes was killed in 212 BCE during the capture of Syracuse by Roman forces under General Marcus Claudius Marcellus in the Second Punic War. Plutarch recounts three slightly different versions of his killing; the most popular one is that Archimedes was contemplating a mathematical diagram when the city was captured. A Roman soldier commanded him to come and meet General Marcellus, but Archimedes declined, saying that he had to finish working on the problem. The famous last words attributed to Archimedes are, "Do not disturb my circles," a reference to the circles in the mathematical drawing that he was supposedly studying.

~

CHAPTER 6

Eratosthenes: Greek (276–194 BCE)

When we look back to the brilliant mathematicians of ancient Greece, our knowledge of their individual lives is, understandably, rather limited. In the case of Eratosthenes, who was largely known for his mathematical and geographical achievements, we again do not have record of many details of his life. Despite this dearth of personal information, we do know that his amazing geographical skills enabled the people of his day to understand and garner an appreciation of the size of the earth. Later we will consider his brilliant method of measurement; but, in the meantime, let us consider his biography—or, at least, what little we know about it.

Eratosthenes was born in 276 BCE in Cyrene, which is now a part of Libya.[1] He was a son of Aglaos and, in his youth, studied in a local school where basic academic subjects were taught. Later, in Athens, he continued his studies, which centered on philosophy. There he also wrote poetry, including *Hermes*, a religiously oriented poem concentrating on the life history of the gods. He also wrote about historical topics, and those writings were quite well received at the time. As a result, in 245 BCE and aged only thirty, he was offered a position as librarian of the Great Library of Alexandria. Upon accepting the role, he moved to Alexandria, where he remained for the rest of his life. Within five years, he was appointed chief librarian, one of the responsibilities of which was tutoring the children of royalty. During his tenure as chief librarian, he enlarged the holdings of the library considerably. He was motivated by the notion that he would want his library to be considered the best in the Greek world. As we can see from his writings, he believed that all humans were good—a theme that contradicted Aristotle, who believed that essentially only the Greeks were good.

In 195 BCE, Eratosthenes contracted ophthalmia, an inflammation of the eye, which resulted in blindness. Afterward, he became very depressed and tried to commit suicide by starving himself. Death actually came a year later, in 194 BCE, when he had reached the rather-old age of eighty-two.

Two contributions have preserved Eratosthenes's name and have allowed him to remain famous today. As we hinted at earlier, Eratosthenes developed a very clever technique for measuring the circumference of the earth. Today, such a task is not terribly difficult; thousands of years ago, though, this was no mean feat. His measuring of the earth was one of the earliest forms of geometry—in fact, the word *geometry* is derived from the Greek for "earth measurement." In about 230 BCE, he measured the earth's circumference, and it was remarkably accurate—less than 2 percent in error.

How did he manage such an accomplishment? To make this measurement, Eratosthenes relied on the relationship of alternate-interior angles of parallel lines, as well as his resources as the chief librarian of Alexandria. Through the library, Eratosthenes had access to records of calendar events. Upon examining these records, he discovered that in a town called Syene (now called Aswan) on the Nile River, the sun was directly overhead

Figure 6.1. Eratosthenes. It is a copper engraving from the 18th century.
See https://www.alamy.com/stock-photo-eratosthenes-of-cyrene-circa
-276-194-bc-greek-scholar-chief-librarian-23533697.html.

at noon on a certain day of the year. As a result of the sun's position, the bottom of a deep well in Syene was entirely lit, and a vertical pole (being parallel to the rays hitting it) cast no shadow.

At the same time, however, a vertical pole in the city of Alexandria did cast a shadow. When that day arrived again, Eratosthenes measured the angle formed by such a pole and the ray of light from the sun that went past the top of the pole to the far end of the shadow ($\angle 1$ in fig. 6.2). He found that angle to be about 7°12', or $\frac{1}{50}$ of 360°.

Assuming the rays of the sun to be parallel, he knew that the angle at the center of the earth must be congruent to $\angle 1$, and, hence, must also measure approximately $\frac{1}{50}$ of 360°. Since Syene and Alexandria were nearly on the same meridian, Syene must be located on the particular radius of the circle that was parallel to the rays of the sun. Eratosthenes thus deduced that the distance between Syene and Alexandria was $\frac{1}{50}$ of the circumference of the earth. The distance from Syene to Alexandria was believed to be about 5,000 Greek stadia. (A stadium was a unit of measurement equal to the length of an Olympic or Egyptian stadium.) Therefore, Eratosthenes concluded that the circumference of the earth was about 250,000 Greek stadia, or about 24,660 miles. This is very close to modern calculations, which have determined the circumference of the earth to be 24,901 miles. So how's that for some real geometry!

Eratosthenes also made a contribution to our understanding of numbers. More specifically, he developed a method that we can use to generate prime numbers—that is, numbers that have exactly two divisors: the number 1 and the number itself. This method uses what we call today the sieve of Eratosthenes, which begins with a table of consecutive numbers (going on as far as you wish) and requires scratching out certain multiples of numbers until all that remain are the prime numbers within the numerical range of the table. Let's consider beginning with the numbers from 1 to 100, as shown in figure 6.3. The procedure that Eratosthenes suggested is to begin with the number 2 and scratch out every multiple of 2 throughout the table. Then, go to the next number that is still not scratched out—the number 3—and once again scratch out all the multiples of that number. Continuing this procedure, we come to the next remaining number, which is 5, and, once again, we scratch out all multiples of 5 in the table. The next number we consider is the number 7, and again we scratch out all of the multiples of 7 that remain on the chart (i.e., the number 49). Using his procedure, we are left with only prime numbers. Therefore, in figure 6.3, we see all the prime

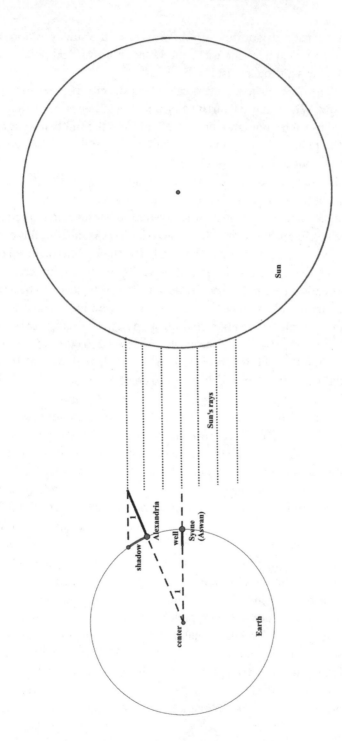

Figure 6.2. Not to scale.

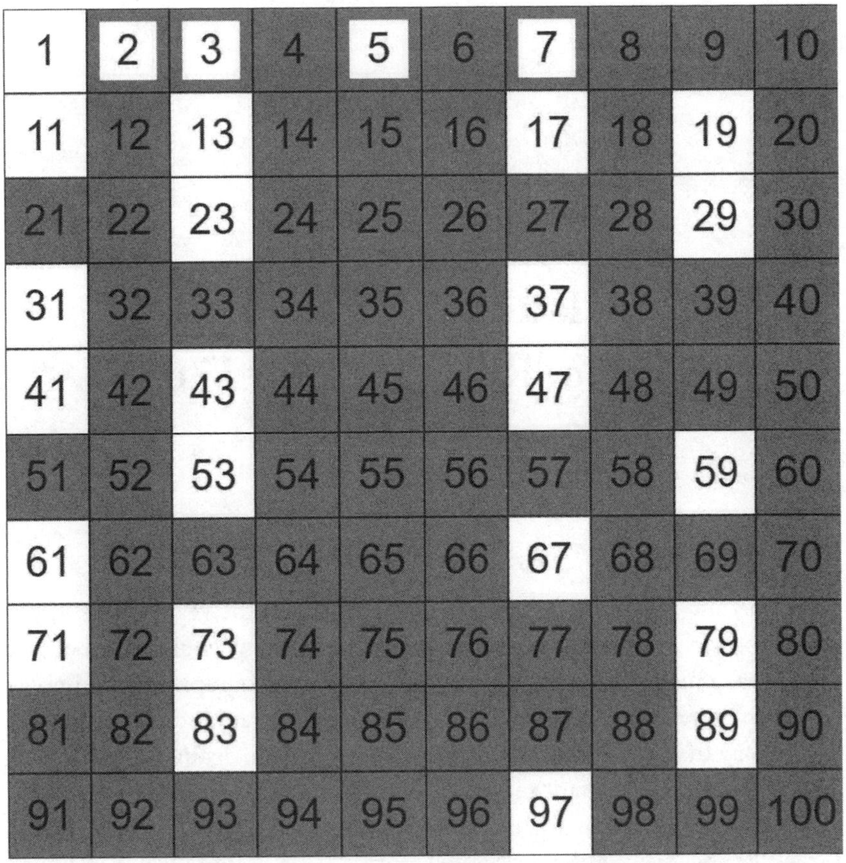

1	2	3	4	5	6	7	8	9	10
11	12	13	14	15	16	17	18	19	20
21	22	23	24	25	26	27	28	29	30
31	32	33	34	35	36	37	38	39	40
41	42	43	44	45	46	47	48	49	50
51	52	53	54	55	56	57	58	59	60
61	62	63	64	65	66	67	68	69	70
71	72	73	74	75	76	77	78	79	80
81	82	83	84	85	86	87	88	89	90
91	92	93	94	95	96	97	98	99	100

Figure 6.3. The sieve of Eratosthenes. In this figure, the numbers 2, 3, 5, and 7 were the only numbers required to eliminate all nonprime numbers between 2 and 100 (that is why they appear in a smaller white box).

numbers up to 100 are those that have not been eliminated. Note: Even though the number 1 was originally on our table, it is by definition *not* a prime number. This is because a prime number has exactly two factors (or divisors): itself and the number 1. Because the number one has only one factor or divisor (the number 1 itself), the number 1 is not a prime number.

Here we have two contributions that Eratosthenes has made to our understanding of mathematics, one in geometry, and the other in number theory. Considering the era in which these discoveries were made, and the fact that they still hold up today, we can say they were quite astonishing.

CHAPTER 7

∽

Claudius Ptolemy:
Greco-Roman (100–170)

Egypt became a Roman province in 30 BCE, and in the year 100 CE, Claudius Ptolemy was born in Alexandria, Egypt. Ptolemy would soon become famous as a mathematician, an astronomer, and a geographer. He is known today for his famous work the *Almagest*, which is an ancient treatise on astronomy that consists of thirteen books. Although, again, we are lacking details about this brilliant mathematician from ancient times, we can deduce some information about his life. Because the *Almagest* was written in Greek, we believe that he descended from a Greek family living in Egypt. We know today that he made all of his astronomical observations from Alexandria between the years 127 and 141 CE.

Ptolemy's theory of planetary motion held that the universe was geocentric, that is, the earth was the center of the universe. This thinking remained intact until the Renaissance, when Nicolaus Copernicus (1473–1543) a Polish astronomer, put forth a heliocentric theory, placing the sun at the center of the universe and indicating that the earth was one of the planets orbiting it. Yet it was the German mathematician and astronomer Johannes Kepler (1571–1630) who, with his three famous laws of elliptical planetary motion, defined the path of these planetary rotations about the sun (see chap. 13).

To do his mathematical calculations of planetary motion, Ptolemy created a table of chords, which was an early form of trigonometric functions, largely equivalent to the sine function. It is perhaps because of his

Figure 7.1. An early Baroque rendition of Claudius Ptolemy
(etching by Theodor de Bry). http://penelope.uchicago.edu/Thayer/E/
Roman/Texts/Ptolemy/Tetrabiblos/home.html.

discoveries related to these chords that we best know of Ptolemy's work. He needed to be able to approximate the lengths of chords in circles in terms of the radius of the circle and the angle cut off by the chord. Therefore, to derive his chord tables, he created a theorem. His chord function was, as mentioned above, related to the sine function:

$$\text{chord}\,\theta = 120\sin\left(\frac{\theta}{2}\right) = 60\left(2\sin\left(\frac{\pi\theta}{360}\text{ radians}\right)\right).$$

But this is not all he was able to do while examining chords: He further came to an approximation of $\pi = 3\frac{17}{120} = 3.141\overline{6}$, by using a regular polygon of 360 sides inscribed in a circle and working with the chords[1]; and he used chord $60° = \sqrt{3} \approx 1.732$.

He is also well known for having established geometric theorems, including a rather-unusual relationship among regular shapes inscribed in a circle. For example, in chapter 10 of book 1 of the *Almagest*, Ptolemy produced geometric theorems that he used to compute the chords discussed above. He claimed that a regular pentagon, hexagon, and decagon—all inscribed in the same circle—would have an unexpected relationship. When the regular pentagon, hexagon, and decagon are inscribed in the circle, the area of the square on one side of the pentagon is equal to the sum of the areas of the squares on one side of the hexagon and on one side of the decagon.

Furthermore, we well know Ptolemy because of the theorem that he developed and that bears his name: Ptolemy's theorem. It states that in a cyclic quadrilateral (i.e., one that is inscribed in a circle), the product of the diagonals is equal to the sum of the products of the opposite sides. The converse of this theorem is also true: if a quadrilateral is such that the product of the diagonals is equal to the sum of the products of the opposite sides, then the quadrilateral can be inscribed in a circle (meaning that each of the four vertices lies on the same circle).

We will explore this in figure 7.2. There we have a quadrilateral $ABCD$ inscribed in the circle O, such that $AC \cdot BD = AB \cdot CD + AD \cdot BC$.

Ptolemy's theorem can only be established with rigid figures, that is, figures whose given information can only describe one figure. For example, a quadrilateral whose side lengths are given can take on various shapes, but if this quadrilateral is inscribed in a circle then it is rigid, since it can only

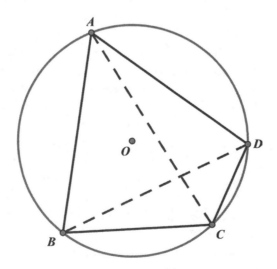

Figure 7.2.

take on one shape. We notice that all triangles are rigid figures; once they are properly defined, they are in a fixed position. In contrast, a quadrilateral is not necessarily in a fixed position, since its shape is not necessarily determined by the lengths of its sides. However, a cyclic quadrilateral *is* a rigid figure, which means that it allows us to establish Ptolemy's theorem. (If the quadrilateral is not cyclic, then the following relationship—sometimes known as the Ptolemy inequality—is as follows: $AC \cdot BD < AB \cdot CD + AD \cdot BC$.)

If we apply Ptolemy's theorem to a rectangle, which is always a cyclic quadrilateral because it can be easily inscribed in a circle, the result is the Pythagorean theorem ($a^2 + b^2 = c^2$). In other words, if we apply Ptolemy's theorem to rectangle *ABCD* shown in figure 7.3. we obtain $dd = ll + ww$, or $d^2 = l^2 + w^2$, which is the Pythagorean theorem as applied to triangle *ABC*.

Another curiosity regarding Ptolemy's theorem can be found by applying it to a regular pentagon inscribed in the circle. Consider the regular pentagon *ABCDE* shown in figure 7.4. Let's apply Ptolemy's theorem to the quadrilateral *ABCD*, noting that all of the sides (*s*) of the pentagon are the same length and the diagonals (*d*) have the same length. Applying Ptolemy's theorem to quadrilateral *ABCD*, we find: $dd = sd + ss$, or $d^2 = sd + s^2$. We now divide through by s^2 to get $\dfrac{d^2}{s^2} = \dfrac{d}{s} + 1$.

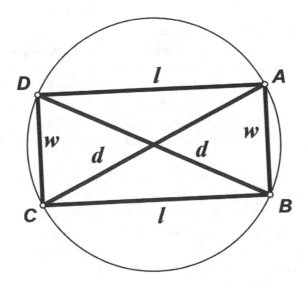

Figure 7.3.

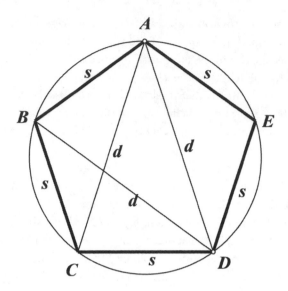

Figure 7.4.

If we let $\dfrac{d}{s} = g$, then we have $g^2 - g - 1 = 0$. Therefore, $g = \dfrac{1+\sqrt{5}}{2} \approx 1.618034$, which is the golden ratio. Interested readers may want to discover the true wonders of the golden ratio, so we recommend *The Glorious Golden Ratio*, by A. S. Posamentier and I. Lehmann.

For its time, another great work by Ptolemy was his eight-book work on geography, which was severely limited in its accuracy because there was little knowledge about the world beyond the Roman Empire and speculation was rampant. In the fifteenth century, a depiction of Ptolemy's understanding of the world map was created, based on his description of the world map and his work in the *Almagest*. That map is presented here in figure 7.5.

Unfortunately, as mentioned above, we know very little about the details of Ptolemy's life, which we believe ended in the year 170 CE in Alexandria, Egypt. Given the disparity between his brilliant geometrical theorems and his remarkable errors and inaccuracy in astronomy and geography, he was undoubtedly an intriguing figure. Here we acknowledge both aspects of his achievements and yet honor him for his famous theorem in geometry, which proves useful to this day.

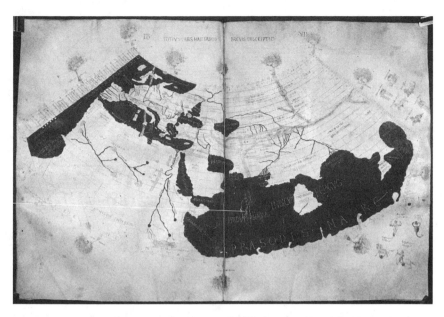

Figure 7.5. Ptolemy's world map, as depicted in the fifteenth century,
based on Ptolemy's works and descriptions. (Francesco di Antonio del Chierico,
1433–1484, an artist of the early renaissance period, in Florence.)

CHAPTER 8

~

Diophantus of Alexandria: Hellenistic Greek (ca. 201–285)

In the study of elementary algebra when one saw an equation that looked like this: $x + y = 7$, the usual reaction was there needs to be a second equation with either x and/or y. Otherwise, the feeling was that a solution could not be found. However, there are, in fact, several solutions to this equation. If we limit our solutions to integral values, one possible solution would be $x = 2$ and $y = 5$, since $2 + 5 = 7$. This type of thinking was first introduced by the Hellenistic Greek mathematician Diophantus, who lived in Alexandria, Egypt, during the middle of the third century of the Common Era and is purported to have lived about eighty-four years. Unfortunately, as with other luminaries from ancient times, very little is known about his life; what has made him well known today is that he is often referred to as the "father of algebra." Although some fragments of his work have been found, his fame today is for a series of books titled *Arithmetica*, which presented algebra for the first time as we know it today, and was a forerunner to the study of number theory.

In *Arithmetica*, Diophantus begins by introducing some concepts of numbers and explains a new notation using a symbol for a variable, something that probably did not catch on for another thousand years, and that today we see as ordinary algebra. He introduced positive and negative numbers, and he was the first to consider fractions as actual numbers. He begins in his first book of *Arithmetica* with some simple problems and then progresses into those that have multiple solutions—albeit in integer form.

Figure 8.1. Diophantus of Alexandria.
https://commons.wikimedia.org/wiki/Category: Diophantus

He later continues to include problems that involve raising numbers to a higher power.

For example, a Diophantine equation—one that may have many solutions, yet all integral—could be the following: $\frac{1}{x}+\frac{1}{y}=\frac{1}{n}$. There are three integer solutions to this equation when $n = 4$:

$$\frac{1}{8}+\frac{1}{8}=\frac{1}{4},$$

$$\frac{1}{6}+\frac{1}{12}=\frac{1}{4}, \text{ and}$$

$$\frac{1}{5}+\frac{1}{20}=\frac{1}{4}.$$

The fame of Diophantus's work has certainly been far-reaching. In 1621, the French mathematician Claude Gaspard Bachet de Méziriac (1581–1638) wrote *Les éléments arithmétiques*, which was a translation from Greek to Latin of Diophantus's *Arithmetica* (see fig. 8.2).

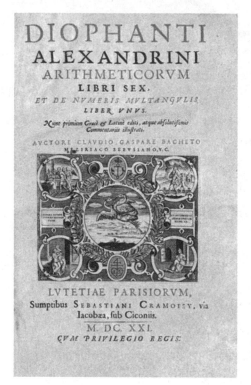

DIOPHANTI
ALEXANDRINI
ARITHMETICORVM
LIBRI SEX.
ET DE NVMERIS MVLTANGVLIS
LIBER VNVS.

*Nunc primùm Græcè & Latinè editi, atque absolutissimis
Commentariis illustrati.*

AVCTORE CLAVDIO GASPARE BACHETO
MEZIRIACO SEBVSIANO, V.C.

LVTETIAE PARISIORVM,
Sumptibus SEBASTIANI CRAMOISY, via
Iacobæa, sub Ciconiis.
M. DC. XXI.
CVM PRIVILEGIO REGIS.

Figure 8.2. Title page of the 1621 edition of Diophantus's *Arithmetica*, translated into Latin by Claude Gaspard Bachet de Méziriac.

A copy of this Bachet translation was owned by the French mathematician Pierre de Fermat (1607–1665). In the margin of a page in the 1621 edition of the book, Fermat wrote that he had a proof that the equation $x^n + y^n = z^n$, where $n \geq 3$, has no integer solutions—but he claimed not to have enough space in the margin to do the proof. Interestingly enough, this statement was included in a 1670 edition of Diophantus's *Arithmetica*, as shown in the next-to-last paragraph in figure 8.3, under the heading *OBSERVATIO DOMINI PETRI DE FERMAT*. This became known as Fermat's last theorem, and it perplexed mathematicians for 358 years, until the British mathematician Andrew Wiles produced a proof in 1995 (see chap. 15).

Diophantine equations do play a role in our everyday computations. Suppose we wish to determine in how many ways you can purchase six-cent stamps and eight-cent stamps for five dollars. Most people will promptly realize that there are two variables that can be represented as x and y. Letting x represent the number of eight-cent stamps and y represent the number of six-cent stamps, the equation $8x + 6y = 500$ should follow. Then, by dividing

Figure 8.3. A page from the 1670 edition of Diophantus's *Arithmetica*, which includes Fermat's commentary, particularly his last theorem.

both sides of the equation by 2, this would be converted to $4x + 3y = 250$. At this juncture, we realize that, although this equation appears to have an infinite number of solutions, it may or may not have an infinite number of *integral* solutions. Moreover, from the context of the original problem, it may or may not have an infinite number of *positive integral* solutions (since a number of stamps must be a positive integer number).

The first consideration is whether integral solutions, in fact, exist. Here we employ a useful theorem, that states that, if the greatest common factor of a and b is also a factor of k, where a, b, and k are integers, then there exist an infinite number of integral solutions (both positive and negative) for x and y in the equation $ax + by = k$. There we have a Diophantine equation, as we are looking for integer solutions.

Since the greatest common factor of 3 and 4 is 1, which is a factor of 250, there exist an infinite number of integral solutions for the equation $4x$

+ 3y = 250. Reverting back to the original problem, we wish to know how many (if any) *positive* integral solutions exist. One possible method developed by Swiss mathematician Leonhard Euler (1707–1783) is often referred to as Euler's method. Following this method, we begin by solving for the variable with the coefficient of least absolute value, which, in this case, is y. Therefore, we get $y = \frac{250-4x}{3}$, which we then write in separate integral parts as $y = 83 + \frac{1}{3} - x - \frac{x}{3} = 83 - x + \frac{1-x}{3}$. We now introduce another variable, t, and let $t = \frac{1-x}{3}$, which allows us to get $x = 1 - 3t$. Since there is no fractional coefficient in this equation, the process does not have to be repeated. Now, substituting x into the original equation, we get $y = \frac{250 - 4(1-3t)}{3} = 82 + 4t$. For various integral values of t, corresponding values of x and y will be generated. From our original problem, we realize that the values of x and y need to be positive integers. Therefore, $x = 1 - 3t > 0$, or $t < \frac{1}{3}$, and $y = 82 + 4t > 0$, which leads to $t > -20\frac{1}{2}$. Combining these gives us $-20\frac{1}{2} < t < \frac{1}{3}$, which tells us that there are 21 possible combinations of six-cent stamps and eight-cent stamps that can be purchased for five dollars.

This is a simple example of how a Diophantine equation can be solved using Euler's method; of course, for this equation, we only considered integral solutions, and, more specifically, only *positive* integral solutions (because of the nature of problem involving the number of stamps purchased). There are many methods to solve Diophantine equations, some of which can take many steps to reach a solution. However, we are indebted to Diophantus of Alexandria for generating these equations and thereby forming the foundation of algebra.

One of the recreational problems in the form of number games in the late fifth century is sometimes considered to be Diophantus's epitaph:

"Here lies Diophantus," the wonder behold.
Through art algebraic, the stone tells how old:
"God gave him his boyhood one-sixth of his life;
One twelfth more as youth while whiskers grew rife;
And then yet one-seventh 'ere marriage begun.
In five years there came a bouncing new son;
Alas, the dear child of master and sage,
After attaining half the measure of his father's life, chill fate took him.
After consoling his fate by the science of numbers for four years, he
    ended his life."[1]

The puzzle implies that Diophantus lived to be about eighty-four years old, which is the only evidence we have of his actual age.

Despite the limited tools that Diophantus had at his disposal, he did solve many mathematical problems and with his work *Arithmetica* inspired mathematicians such as the Arabic mathematician al-Karaji (c.980–1030) and also set the stage for the French mathematician Pierre de Fermat (1601–1665), who might be considered the founder of modern number theory.

CHAPTER 9

~

Brahmagupta:
Indian (598–668)

It is well known that our current system of numbers stems originally from India; yet it reached Western Europe via the Arabs, with whom Fibonacci worked in the early part of the thirteenth century (see chap. 10). Hence, we call our number system the Hindu-Arabic numerals. The mathematician perhaps most responsible for spreading this Indian-numeral system is the mathematician Brahmagupta, who was born in India in the year 598 CE in the town of Bhillamala (today, Bhinmal), which was the capital of Gurjaradesa, the second-largest kingdom of Western India. Brahmagupta was largely interested in astronomy, but also showed a great deal of creativity in mathematics, and it is for this reason that here we will highlight some of his findings. However, before we move on to his mathematical accomplishments, we should mention some of his astronomical discoveries, which include establishing that the earth is closer to the moon than it is to the sun, and calculating the earth's circumference to be about 22,500 miles. (The actual circumference of the earth is 24,901 miles.) He also found that the length of a year was 365 days, 6 hours, 12 minutes, and 19 seconds, which is close to the actual year length that we know today: 365 days, 5 hours, 48 minutes, and 45 seconds.

In the year 628 CE, Brahmagupta wrote a book called *Brāhmasphuṭasiddhānta* (*Brahma's Correct System of Astronomy*), which was based on previous works but also contained many of his new ideas, some of which we will present later. One striking feature of his book is that it was the first

Figure 9.1. Brahmagupta. Nineteenth-century illustration of a Hindu astronomer. Original caption: "Dybuck, an astronomer, calculating an Eclipse." The illustration, as well as the term *dybuck*, is derived from an etching with the title "Daybouk ou astronome hindou" by Frans Balthazar Solvyns (between 1791 and 1803), published in his *Les Hindous* (1808). See https://gl.wikipedia.org/wiki/Brahmagupta#/media/File:Hindu_astronomer,_19th-century_illustration.jpg.

to mention zero as a number rather than as a placeholder. Later in *Brāhmasphuṭasiddhānta*, he describes arithmetic operations on negative numbers. He even delved into the question of zero divided by zero, which he defined as equal to zero; as we know today, though, that division remains as undefined.

Brāhmasphuṭasiddhānta contains twenty-four chapters; chapter 18, on calculations, presents a method of finding both the square and the square root of numbers, as well as the cube and the cube root of numbers. He also introduced fractions in the way we write them today, and he provided the arithmetic for adding and multiplying fractions; for example,

$$\frac{a}{c} + \frac{b}{d} \cdot \frac{a}{c} = \frac{a(d+b)}{cd}.$$

He also presented a formula for finding the solution to a linear equation, then leads up to the solution of a quadratic equation. Thus, he provided us with a variation of the formula that we are accustomed to using for solving the quadratic equation $ax^2 + bx + c = 0$, namely, $x = \frac{-b \pm \sqrt{b^2 - 4ac}}{2a}$.

Brahmagupta also provided formulas for finding the sum of the squares of the first n natural numbers, $\frac{n(n+1)(2n+1)}{6}$, and the cubes of the first n natural numbers, $\left(\frac{n(n+1)}{2}\right)^2$. He also developed a way of generating Pythagorean triples by letting $a = mx$, $b = m + d$, and $c = m(1 + x) - d$, where $d = \frac{mx}{x+2}$; with this we can then show by means of a simple algebraic application, $a^2 + b^2 = c^2$.

Perhaps the relationship that Brahmagupta is best known for is the formula he developed for finding the area of a cyclic quadrilateral, or a quadrilateral for which all four vertices lie on the same circle. Referring to the quadrilateral $ABCD$ in figure 9.2, where the lengths of the sides are marked as a, b, c, and d, Brahmagupta showed that the area of the cyclic quadrilateral $ABCD$ can be found by the formula $\sqrt{(s-a)(s-b)(s-c)(s-d)}$, where s is the semiperimeter, that is, $s = \frac{a+b+c+d}{2}$.

This is an interesting extension of the famous formula that the Roman mathematician Hero of Alexandria (10–70 CE) developed for finding the area of a triangle, given only the lengths of the sides, a, b, and c: Area = $\sqrt{s(s-a)(s-b)(s-c)}$, where, once again, s is the semiperimeter. In effect, Brahmagupta considered Hero's formula as treating the triangle as if it were a quadrilateral with the side $d = 0$.

An interesting extension of Brahmagupta's formula to the general quadrilateral is that the area of any (convex) quadrilateral = $\sqrt{(s-a)(s-b)(s-c)(s-d) - abcd \cdot \cos^2\left(\frac{a+\gamma}{2}\right)}$, where, once again, a, b, c, and d are the

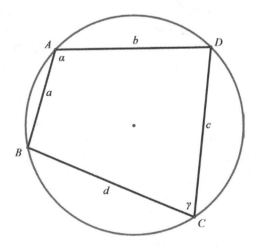

Figure 9.2.

lengths of the sides, $s = \frac{a+b+c+d}{2}$, and α and γ are the measures of a pair of opposite angles of the quadrilateral.

This formula shows that of all quadrilaterals that can be formed from four given side lengths, the one with the maximum area is the cyclic quadrilateral. The maximum area is achieved when $abcd \cdot \cos^2\left(\frac{\alpha+\gamma}{2}\right) = 0$, which occurs when α + γ = 180°—a fact that holds true only for cyclic quadrilaterals.

Brahmagupta also found that for a cyclic quadrilateral of consecutive sides of lengths a, b, c, and d, where m and n are the lengths of the diagonals, the following relationship holds true:

$$m^2 = \frac{(ab+cd)(ac+bd)}{ad+bc}$$
$$n^2 = \frac{(ac+bd)(ad+bc)}{ab+cd}.$$

Another interesting relationship regarding cyclic quadrilaterals and attributed to Brahmagupta is that in a cyclic quadrilateral with perpendicular diagonals, the line through the point of intersection of the diagonals and perpendicular to a side of the quadrilateral bisects the opposite side.

The proof of this is rather simple and gives a further insight into cyclic quadrilaterals. Consider figure 9.3, where diagonals AC and BD of cyclic quadrilateral ABCD are perpendicular at G, and GE ⊥ AED. We

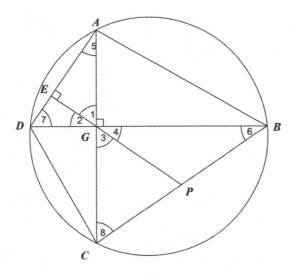

Figure 9.3.

just need to prove that *GE* bisects *BC* at *P*. In right triangle *AEG*, $\angle 5$ is complementary to $\angle 1$, and $\angle 2$ is complementary to $\angle 1$. Therefore, $\angle 5 = \angle 2$. However, $\angle 2 = \angle 4$. Thus, $\angle 5 = \angle 4$. Since $\angle 5$ and $\angle 6$ are equal in measure to half the measure of arc *DC*, they are congruent. Therefore, $\angle 4 = \angle 6$, and *BP* = *GP*. Similarly, $\angle 7 = \angle 3$ and $\angle 7 = \angle 8$, so that *GP* = *PC*. Thus *CP* = *PB*.

Using his mathematical talents, Brahmagupta was also a major player in the development of astronomy, as evidenced by an astronomical treatise he published in 667, which is entitled *Khaṇḍakhādyaka*. Thus, Brahmagupta is known in the world of astronomy as well as mathematics. He died shortly thereafter, in the year 668, in the town of Ujjain, India. His legacy today is primarily extending Hero's formula for the area of a triangle to that of a cyclic quadrilateral.

CHAPTER 10

~

Leonardo Pisano Bigollo, "Fibonacci": Italian (1170–1250)

With the dawn of the thirteenth century, both the field of mathematics and the European world began to acquire their modern image. This was largely due to the Italian mathematician Leonardo Pisano Bigollo, best known as Fibonacci. Fibonacci forever changed Western methods of calculation, which facilitated the exchange of currency and trade. Furthermore, he presented mathematicians with challenges that remain unsolved to this day; they are published in countless books and provide material for a journal published quarterly since 1963 by the Fibonacci Association.

Leonardo Pisano Bigollo, or Leonardo of Pisa, is today known as Fibonacci. The name Fibonacci possibly derived from the Latin *filius Bonacci*, meaning a son of Bonacci, but, more likely, it might have been derived from *de filiis Bonacci*, referring to the family of Bonacci. He was born to the wealthy Italian merchant Guglielmo Bonacci and his wife in the port city of Pisa, Italy, around 1170 shortly after the start of construction of the famous bell tower known today as the Leaning Tower of Pisa. These were turbulent times in Europe. The Crusades were in full swing, and the Holy Roman Empire was in conflict with the papacy. The cities of Pisa, Genoa, Venice, and Amalfi, although frequently at war with each other, were maritime republics with specified trade routes to the Mediterranean countries and beyond. Pisa had played a powerful role in commerce since Roman times, and, even earlier, it served as a port of call for Greek traders. Early

Figure 10.1. Fibonacci.

on, Pisa had established outposts for its commerce among its colonies and along trading routes.

In 1192, Guglielmo Bonacci became a public clerk in the customs house for the Republic of Pisa, which was stationed in the Pisan colony of Bugia (today Bejaia, Algeria) on the Barbary Coast of Africa. Shortly after his arrival, he brought his son, Leonardo, to join him so that the boy could learn the skill of calculating and become a merchant. The ability to perform calculations was significant, since each republic had its own units of money and traders had to calculate monies due them. This entailed determining currency equivalents on a daily basis. It was in Bugia that Fibonacci first became acquainted with the "nine Indian figures," as he called the Hindu numerals, and "the sign 0 which the Arabs call zephyr." He declares his fascination for the methods of calculation using these numerals in the only source we have about his life story, the prologue to his most famous book, *Liber Abaci* (*The Book of Calculation*), which he wrote in 1202 and revised in 1228 (see fig. 10.2). This was the first time Hindu-Arabic numerals appeared in Europe. During his time away from Pisa, he received instruction from a Muslim teacher who introduced him to a book on algebra titled *al-Kitāb al-mukhtaṣar fī ḥisāb al-jabr wal-muqābala* (*The Compendious Book on Calculation by Completion and Balancing*) by the

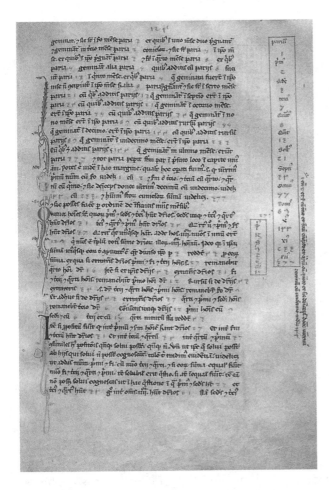

Figure 10.2. A page of Fibonacci's *Liber Abaci* (from the Biblioteca Nazionale di Firenze), showing the Fibonacci numbers in the right margin.

Persian mathematician Muḥammad ibn Mūsā al-Khwārizmī (ca. 780–ca. 850), which greatly influenced him. By the way, the name "algebra" comes from the title of this book.

During his lifetime, Fibonacci traveled extensively to Egypt, Syria, Greece, Sicily, and Provence, where he not only conducted business but also met with mathematicians to learn their ways of doing mathematics.. When he returned to Pisa around the turn of the century, Fibonacci began to write about calculation methods with the Indian numerals for commercial applications in his book *Liber Abaci*. The volume consists largely of algebraic problems of "real world" situations that require more-abstract mathematics. Fibonacci wanted to spread these newfound techniques to his compatriots.

Bear in mind that during these times, the printing press had not yet been invented, so books had to be handwritten by scribes; if a copy was to be made, that, too, had to be handwritten. Fibonacci had written other works, such as *Practica Geometriae* (1220), a book on the practice of geometry. It covers geometry and trigonometry with a rigor comparable to that of Euclid with ideas presented in proof form as well as in numerical form, using these "new," very convenient, numerals. Here, Fibonacci uses algebraic methods to solve geometric problems, as well as the reverse. In 1225, he wrote *Flos* (on flowers or blossoms) and *Liber quadratorum* (*The Book of Squares*), the latter of which truly distinguished Fibonacci as a talented mathematician, and ranking him very high among number theorists. Fibonacci likely wrote additional works; however, there is no trace of them today. His book on commercial arithmetic, *Di minor guisa*, is lost, as is his *Commentary on Book X* of Euclid's *Elements*, which contained a numerical treatment of irrational numbers, as compared to Euclid's geometrical treatment of them.

The confluence of politics and scholarship brought Fibonacci into contact with the Holy Roman Emperor Frederick II (1194–1250) in the third decade of the century. Frederick had spent the years up to 1227 consolidating his power in Italy; he had been crowned king of Sicily in 1198, then king of Germany in 1212, and then, by the pope in St. Peter's Cathedral in Rome, Holy Roman Emperor in 1220. In its maritime conflicts with Genoa and its land-based conflicts with Lucca and Florence, Frederick supported Pisa, which then had a population of about ten thousand. As a strong patron of science and the arts, Frederick became aware of Fibonacci's work through the scholars at his court who had corresponded with Fibonacci since his return to Pisa around 1200. These scholars included Michael Scotus, who was the court astrologer, and the person to whom Fibonacci dedicated his book *Liber Abaci*; Theodorus Physicus, the court philosopher; and Dominicus Hispanus, who, when Frederick's court met in Pisa around 1225, suggested to Frederick that he meet Fibonacci. The meeting took place as expected within the year.

Johannes of Palermo, another member of Frederick II's court, presented a number of problems as challenges to the great mathematician Fibonacci. He solved three of these problems, the solutions for which he provided in *Flos*, which he sent to Frederick II. One of the problems he was able to solve, which was taken from the Persian mathematician Omar Khayyam's (1048–1131) book on algebra, was to solve the equation: $x^3 + 2x^2 + 10x = 20$.

Fibonacci knew that this was not solvable with the numerical system then in place—the Roman numerals. He provided an approximate answer, pointing out that the answer was neither an integer, nor a fraction, nor the square root of a fraction. Without any explanation, he gave his approximate solution in the form of a sexagesimal number (i.e., a number from a base-60 numerical system): 1.22.7.42.33.4.40, which is equal to $1+\frac{22}{60}+\frac{7}{60^2}+\frac{42}{60^3}+\frac{33}{60^4}+\frac{4}{60^5}+\frac{40}{60^6}$. However, with today's computer-algebra system, we can identify the proper solution—which is by no means trivial! It is

$$x = -\sqrt[3]{\frac{2\sqrt{3930}}{9} - \frac{352}{27}} + \sqrt[3]{\frac{2\sqrt{3930}}{9} + \frac{352}{27}} - \frac{2}{3} \approx 1.3688081075,$$

which compares to Fibonacci's value of 1.3924... .

Another of the problems with which he was challenged and was able to solve is one we can explore here, since it doesn't require anything more than some knowledge of basic algebra. Remember that although these methods may seem elementary to us, they were hardly known at the time of Fibonacci, and so this was considered a real challenge. The problem was to find the perfect square that remains a perfect square when increased or decreased by 5.

Fibonacci found the number $\frac{41}{12}$ as his solution to the problem. To check this, we must both add 5 to and subtract 5 from the number, then see if the result is still a perfect square:

$$\left(\frac{41}{12}\right)^2 + 5 = \frac{1681}{144} + \frac{720}{144} = \frac{2401}{144} = \left(\frac{49}{12}\right)^2$$

$$\left(\frac{41}{12}\right)^2 - 5 = \frac{1681}{144} - \frac{720}{144} = \frac{961}{144} = \left(\frac{31}{12}\right)^2$$

Since both results from the addition and subtraction are perfect squares, we have shown that $\frac{41}{12}$ meets the criteria set out in the problem. Luckily, the problem asked for 5 to be added and subtracted from the perfect square; had he been asked to add or subtract 1, 2, 3, or 4 instead of 5, the problem could not have been solved.

The third problem, whose solution Fibonacci also presented in *Flos*, was to solve the following: Three men are to share an amount of money in the following parts: $\frac{1}{2}$, $\frac{1}{3}$, and $\frac{1}{6}$. Each person takes some money from this total amount until there is nothing left. The first man then returns $\frac{1}{2}$ of what he took; the second, $\frac{1}{3}$ of what he took; and the third, $\frac{1}{6}$ of what he took. When the total of what was returned is divided equally among the three, each has his correct share, namely, $\frac{1}{2}$, $\frac{1}{3}$, and $\frac{1}{6}$. What was the original amount of money, and how much did each person get from that original sum?

Although none of Fibonacci's competitors could solve any of these three problems, to the final question he determined that 47 was the smallest amount possible for the original sum of money, but he claimed that the problem was indeterminate.

In 1240, Fibonacci was honored with a lifetime salary by the Republic of Pisa for his service to the people, whom he advised on matters of accounting, often pro bono. We do not know exactly when Fibonacci died, but it is believed that he died in Pisa at some point between 1240 and 1250.

Although Fibonacci was considered one of the greatest mathematicians of his time, his fame today is largely based on the book, *Liber Abaci*. To appreciate his work, let us consider his most well-known text as an example. This extensive volume is full of very interesting problems. *Liber Abaci* was based on the knowledge of arithmetic and algebra that Fibonacci had accumulated during his travels; furthermore, it was widely copied and imitated, and, as we noted above, it introduced to Europe both the Hindu-Arabic place-valued decimal system and Arabic numerals. The book was increasingly widely used for the better part of the next two centuries—a bestseller! Fibonacci begins his famous book *Liber Abaci* with the following:

The nine Indian figures are:

9 8 7 6 5 4 3 2 1.

With these nine figures, and with the sign 0, which the Arabs call zephyr, any number whatsoever is written, as demonstrated below. A number is a sum of units, and through the addition of them the number increases by steps without end. First one composes those numbers, which are from one to ten. Second, from the tens are made those numbers, which are from ten up to one hundred. Third, from the hundreds are made those numbers, which are from one hundred up to one thousand. . . . and thus, by an unending sequence of steps, any number whatsoever is constructed by joining the preceding numbers. The first place in the writing of the numbers is at the right. The second follows the first to the left.

Fibonacci used the term "Indian figures" to refer to the Hindu numerals. Despite their relative facility, these numerals were not widely accepted by merchants, who were suspicious of any who knew how to use them. These merchants were simply afraid of being cheated. We can safely say that it took the same three hundred years for these numerals to catch on as it did for the leaning tower of Pisa to be completed.

Interestingly, *Liber Abaci* also contains simultaneous linear equations. Many of the problems that Fibonacci considers, however, were similar to those appearing in Arab sources. This does not detract from the value of the book, since it is the collection of the *solutions* to these problems that establishes *Liber Abaci* as a major contribution to our development of mathematics. As a matter of fact, a number of mathematical terms that are common today were first introduced in Fibonacci's most famous text. Within it, he referred to "*factus ex multiplicatione*"; this is our first record of these words, from which we now speak of the "factors of a multiplication." Incidentally, two other words whose introduction into the current mathematics vocabulary seems to stem from this famous book are "numerator" and "denominator."

The second section of *Liber Abaci* includes a large collection of problems aimed at merchants. They relate to the price of goods, how to convert between the various currencies in use in Mediterranean countries, calculate profit on transactions, and problems that had probably originated in China.

Fibonacci was aware of a merchant's desire to circumvent the church's ban on charging interest on loans. Therefore, he devised a way to hide the interest in a higher initial sum than the actual loan, and then base the calculations on compound interest.

The third section of the book contains many problems such as:

A hound whose speed increases arithmetically chases a hare whose speed also increases arithmetically, how far do they travel before the hound catches the hare?.

A spider climbs so many feet up a wall each day and slips back a fixed number each night, how many days does it take him to climb the wall?.

Calculate the amount of money two people have after a certain amount changes hands and the proportional increase and decrease are given.

There are also problems involving perfect numbers (those numbers for which the sum of their proper factors is equal to the number itself), there are also problems where Fibonacci employs to the Chinese remainder theorem, which states that if one knows the remainders of the division of a number by various integers than one could also determine the remainder of a division by the product of these integers, assuming that the devices are relatively prime. Once again, this well preceded a formal study of number theory. He also introduces problems involving the sums of arithmetic and geometric series. Fibonacci treats numbers such as $\sqrt{15}$ in the fourth section, both with rational approximations and with geometric constructions. This treatment of an irrational number was not really studied until centuries later, so we might say that Fibonacci was well ahead of his time!

Some of the classical problems, which are considered recreational mathematics today, first appeared in the Western world in *Liber Abaci*. This book is of particular interest to us because it was the first publication in Western culture to use the Hindu numerals to replace the clumsy Roman

Beginning	1	"A certain man had one pair of rabbits together in a certain enclosed place, and one wishes to know how many are created from the pair in one year when it is the nature of them in a single month to bear another pair, and in the second month those born to bear also. Because the above written pair in the first month bore, you will double it; there will be two pairs in one month. One of these, namely the first, bears in the second month, and thus there are in the second month 3 pairs; of these in one month two are pregnant and in the third month 2 pairs of rabbits are born and thus there are 5 pairs in the month; in this month 3 pairs are pregnant and in the fourth month there are 8 pairs, of which 5 pairs bear another 5 pairs; these are added to the 8 pairs making 13 pairs in the fifth month; these 5 pairs that are born in this month do not mate in this month, but another 8 pairs are pregnant, and thus there are in the sixth month 21 pairs; to these are added the 13 pairs that are born in the seventh month; there will be 34 pairs in this month; to this are added the 21 pairs that are born in the eighth month; there will be 55 pairs in this month; to these are added the 34 pairs that are born in the ninth month; there will be 89 pairs in this month; to these are added again the 55 pairs that are both in the tenth month; there will be 144 pairs in this month; to these are added again the 89 pairs that are born in the eleventh month; there will be 233 pairs in this month. To these are still added the 144 pairs that are born in the last month; there will be 377 pairs and this many pairs are produced from the above-written pair in the mentioned place at the end of one year.
First	2	
Second	3	
Third	5	
Fourth	8	
Fifth	13	
Sixth	21	
Seventh	34	
Eighth	55	
Ninth	89	
Tenth	144	You can indeed see in the margin how we operated, namely that we added the first number to the second, namely the 1 to the 2, and the second to the third and the third to the fourth and the fourth to the fifth, and thus one after another until we added the tenth to the eleventh, namely the 144 to the 233, and we had the above-written sum of rabbits, namely 377 and thus you can in order find it for an unending number of months."
Eleventh	233	
Twelfth	377	

Figure 10.3. The rabbit problem, as it was stated
(with the left-marginal note included).

numerals; because Fibonacci was the first to use a horizontal fraction bar; and because it casually includes a recreational mathematics problem that has made Fibonacci famous for posterity. This is the problem of the regeneration of rabbits (see fig. 10.3).

To see how this problem's situation would look on a monthly basis, consider the chart in figure 10.4. If we assume that a pair of baby (*B*) rabbits matures in one month to become offspring-producing adults (*A*), then we can set up the following chart:

This problem generated the sequence of numbers

1, 1, 2, 3, 5, 8, 13, 21, 34, 55, 89, 144, 233, 377, . . .

Today, this sequence is known as the *Fibonacci numbers*. At first glance, there is nothing spectacular about these numbers beyond the relationship that would allow us to generate additional numbers of the sequence quite easily. We notice that every number in the sequence (after the first two) is the sum of the two preceding numbers. The Fibonacci sequence can be written in a way so that its recursive definition becomes clear: each number is the sum of the two preceding ones:

1
  1
$$1 + 1 = 2$$
$$1 + 2 = 3$$
$$2 + 3 = 5$$
$$3 + 5 = 8$$
$$5 + 8 = 13$$
$$8 + 13 = 21$$
$$13 + 21 = 34$$
$$21 + 34 = 55$$
$$34 + 55 = 89$$
$$55 + 89 = 144$$
$$89 + 144 = 233$$
$$144 + 233 = 377$$
$$233 + 377 = 610$$
$$377 + 610 = 987$$
$$610 + 987 = 159$$

Figure 10.4.

The Fibonacci sequence is the oldest known (recursive) *recurrent* sequence. Although there is no direct evidence that Fibonacci knew of this relationship, we can safely assume that a man of his talents and insight would have recognized it. It took another four hundred years before this relationship appeared in print beyond his book.

These numbers were not identified as anything special during the time Fibonacci wrote *Liber Abaci*. As a matter of fact, the famous German mathematician and astronomer, Johannes Kepler (1571–1630), mentioned these numbers in a 1611 publication[1] when he said that the ratios "as 5 is to 8, so is 8 to 13, so is 13 to 21 almost." Centuries passed and the numbers still went unnoticed. In the 1830s, C. F. Schimper and A. Braun noticed the numbers appeared as the number of spirals of bracts on a pinecone. In the mid-1800s the Fibonacci numbers began to capture the fascination of mathematicians. They took on their current name ("Fibonacci numbers") from François-Édouard-Anatole Lucas (1842–1891), the French mathematician usually referred to as "Edouard Lucas," who later devised his own sequence by following the pattern set by Fibonacci. Lucas numbers form a sequence of numbers much like the Fibonacci numbers, and also closely related to the

Figure 10.5. François-Édouard-Anatole Lucas.

Fibonacci numbers. Instead of starting with 1, 1, 2, 3, 5, 8, 13, 21, . . . , Lucas began his sequence with 1, 3, 4, 7, 11, 18, 29, ...

At about this time the French mathematician, Jacques-Philippe-Marie Binet (1786–1856), developed a formula for finding any Fibonacci number given its position in the sequence. That is, with Binet's formula we can find the 118th Fibonacci number without having to list the previous 117 numbers. The formula is:

$$F_n = \frac{1}{\sqrt{5}}\left[\left(\frac{1+\sqrt{5}}{2}\right)^n - \left(\frac{1-\sqrt{5}}{2}\right)^n\right],$$

where F_n is the nth Fibonacci number. The Fibonacci numbers are probably the most famous and ubiquitous sequence of numbers in all of mathematics, since they appear in just about every aspect of our experiences.

Still one may ask, what is so special about these numbers? Let us just begin to scratch the surface by simply inspecting this famous Fibonacci number sequence and some of the remarkable properties it has.

As we did above, we will use the symbol F_7 to represent the 7th Fibonacci number, and F_n to represent the nth Fibonacci number. Consider the first 30 Fibonacci numbers shown in figure 10.6.

With the seemingly endless applications of these lovely Fibonacci numbers, there must be a simple way to get the sum of a specified number of them. A simple formula would be helpful, as opposed to actually adding all the Fibonacci numbers to a certain point. To derive such a formula for the sum of the first n Fibonacci numbers, we will use a technique that will help us generate a formula. From the definition of the Fibonacci numbers, we can write that symbolically as $F_{n+2} = F_{n+1} + F_n$ where $n > 1$. This can be rewritten as $F_n = F_{n+2} - F_{n+1}$. By substituting increasing values for n we get the following:

$$F_1 = F_3 - F_2$$
$$F_2 = F_4 - F_3$$
$$F_3 = F_5 - F_4$$
$$F_4 = F_6 - F_5$$
$$\vdots$$
$$F_{n-1} = F_{n+1} - F_n$$
$$F_n = F_{n+2} - F_{n+1}$$

By adding these equations, you will notice that there will be many terms on the right side of the equations that will disappear (because their

$$F_1 = 1$$
$$F_2 = 1$$
$$F_3 = 2$$
$$F_4 = 3$$
$$F_5 = 5$$
$$F_6 = 8$$
$$F_7 = 13$$
$$F_8 = 21$$
$$F_9 = 34$$
$$F_{10} = 55$$
$$F_{11} = 89$$
$$F_{12} = 144$$
$$F_{13} = 233$$
$$F_{14} = 377$$
$$F_{15} = 610$$
$$F_{16} = 987$$
$$F_{17} = 1597$$
$$F_{18} = 2584$$
$$F_{19} = 4181$$
$$F_{20} = 6765$$
$$F_{21} = 10946$$
$$F_{22} = 17711$$
$$F_{23} = 28657$$
$$F_{24} = 46368$$
$$F_{25} = 75025$$
$$F_{26} = 121393$$
$$F_{27} = 196418$$
$$F_{28} = 317811$$
$$F_{29} = 514229$$
$$F_{30} = 832040$$

Figure 10.6.

sum is zero—since you will be adding and subtracting the same number). What will remain on the right side will be $F_{n+2} - F_2 = F_{n+2} - 1$.

On the left side we have the sum of the first n Fibonacci numbers: $F_1 + F_2 + F_3 + F_4 + \cdots + F_n$, which is what we are looking for. Therefore, we get the following: $F_1 + F_2 + F_3 + F_4 + \cdots + F_n = F_{n+2} - 1$, which says that the sum of the first n Fibonacci numbers is equal to the Fibonacci number two further along the sequence minus 1. This can also be written symbolically as

$$\sum_{i=1}^{n} F_i = F_{n+2} - 1 \cdot$$

Just for entertainment, and to entice you a bit to perhaps enjoy the Fibonacci numbers in greater detail consider the following illustrations.

The sum of *any ten* consecutive Fibonacci numbers is divisible by 11. We could convince ourselves that this may be true by considering some randomly chosen examples. Take, for example, the sum of the following ten consecutive Fibonacci numbers: $13 + 21 + 34 + 55 + 89 + 144 + 233 + 377 + 610 + 987 = 2{,}563$, which is divisible by 11, since $11 \cdot 233 = 2{,}563$. We could repeat this for any other sum of 10 consecutive Fibonacci numbers, such as the sum of the ten consecutive Fibonacci numbers from F_{21} to F_{30}, which is $2{,}160{,}598 = 11 \cdot 196{,}418$. One way to go about convincing yourself of the truth in this "conjecture" is to keep on taking the sum of groups of ten consecutive Fibonacci numbers and checking to see if the sum is a multiple of 11. You could also try to prove the statement, mathematically. Listing the remainders of the first few Fibonacci numbers upon dividing by 11, we have

1, 1, 2, 3, 5, 8, 2, 10, 1, 0, <u>1, 1, 2, 3, 5, 8, 2, 10, 1, 0</u>, . . .

We see that the remainders repeat in cycles of length 10. Since it is the remainder upon dividing a number by 11 that determines its divisibility by 11, all we have to do is check that in adding any 10 consecutive numbers in the sequence 1, 1, 2, 3, 5, 8, 2, 10, 1, 0, 1, 1, 2, 3, 5, 8, 2, 10, 1, 0, . . . we get a sum divisible by 11. We can check this as follows. Since the cycle of this sequence is of length exactly 10, adding any 10 consecutive numbers in this sequence will always come out to adding the 10 numbers—1, 1, 2, 3, 5, 8, 2, 10, 1, 0—in a cycle.

Imagine these 10 numbers arranged in a clockwise order around a circle (fig. 10.7), with the sequence above obtained by traveling around the circle over and over. Then you can see that any numbers missed at the beginning of a cycle—if the sum is started somewhere in the interior of the cycle—are regained from the next cycle; for example, the sum 5 + 8 + 2 + 10

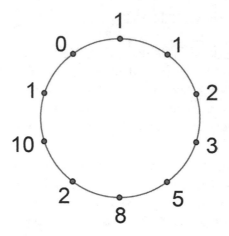

Figure 10.7.

+ 1 + 0 + <u>1 + 1 + 2 + 3</u>. This is because no matter where you start counting on the circle, counting 10 numbers clockwise around the circle will amount to counting all 10 numbers, because that's exactly how many numbers there are. These 10 numbers have sum 33, which is indeed divisible by 11. (Also see: *The Fabulous Fibonacci Numbers*, A. S. Posamentier and I. Lehmann, Prometheus Books, 2007).

    Aside from the fact that Fibonacci is largely responsible for the use of our decimal number system and the numerical symbols we use, having introduced them to the Western world in 1202, he is primarily remembered today for the ubiquitous numbers that bear his name—the Fibonacci numbers.

CHAPTER 11

Gerolamo Cardano: Italian (1501–1576)

"You know what the fellow said—in Italy, for thirty years under the Borgias, they had warfare, terror, murder, and bloodshed, but they produced Michelangelo, Leonardo da Vinci, and the Renaissance. In Switzerland, they had brotherly love, they had five hundred years of democracy and peace— and what did that produce? The cuckoo clock." These lines were spoken by Orson Welles, playing Harry Lime, the villain in the 1949 British film noir *The Third Man*, directed by Carol Reed (1906–1976). The English novelist Graham Greene (1904–1991) wrote the screenplay, but he credited these lines to Orson Welles, who probably added them when some extra dialogue was needed while the film was being shot.[1] Although not historically accurate and dramatically exaggerated, the drawn comparison contains a grain of truth that cannot be denied. Without the patronage of tyrants who had plenty of money at their disposal, the Italian Renaissance would not have brought about so many masterpieces of architecture, sculpture, and painting. Democratic structures, on the other hand, seem to be less generous when it comes to financing arts and culture. The House of Borgia was an Italo-Spanish noble family, which became very powerful in the fifteenth and sixteenth centuries. The Borgias were involved in many ecclesiastical and political affairs and wholly without scruple in the choice of means to increase their power and influence. Like other wealthy dynasties such as the Medici, they dominated local governments and, over several generations, extended their political influence over wider parts of Italy and Europe. In the second half of the fifteenth century, they even gained control over the papacy, and the Borgia produced two popes: Alfonso de Borgia, who ruled as Pope Callixtus III during 1455–1458, and Rodrigo Lanzol Borgia, as Pope Alexander VI, during 1492–1503. Alexander VI was one of the most memorable of the corrupt and secular popes of the Renaissance; the Borgias were suspected of many crimes during his reign—including murder. In the course of his pontificate, Alexander appointed forty-seven cardinals to further his political policies. He had several illegitimate children, and in 1493 he made his teenaged son Cesare a cardinal. The name Borgia became a byword for libertinism and nepotism. Incidentally, in the same year, 1493, Pope Alexander followed a request by Queen Isabella and King Ferdinand of Spain and issued a papal bull granting Spain the exclusive right to claim the New World lands discovered by Christopher Columbus. However, in spite of their cold-blooded greed for power, the Borgias as well as the Medici

Figure 11.1. Gerolamo Cardano. Copper engraving from the seventeenth century, artist unknown. https://commons.wikimedia.org/wiki/File:Jer%C3%B4me_Cardan.jpg

were also great patrons of the arts who contributed significantly to the Renaissance. In particular, they brought the spirit of Renaissance art and philosophy into the Vatican. Alexander had the University of Rome rebuilt and hired the greatest professors to teach there. He restored and embellished the Vatican palaces and persuaded Michelangelo to draw plans for the rebuilding of St. Peter's Basilica. In 1502, his son Cesare, by now commander of the papal armies, hired Leonardo da Vinci as his chief military engineer and architect. The Italian Renaissance began during the fourteenth century, and it was the earliest manifestation of the general European Renaissance, a period of great cultural change and achievement that marked the transition between medieval and modern Europe. While the most famous figures of the Italian Renaissance are artists such as Leonardo da Vinci, Michelangelo, or Raphael, great advances also occurred in mathematics, contributing substantially to the transition from natural philosophy to modern science and the scientific revolution in which Galileo Galilei (1564–1642) would play a central role.

Perhaps the most influential mathematician of the Italian Renaissance was Gerolamo Cardano, and his biography fits quite perfectly to Orson Welles's description of Italy during the reign of the Borgias.[2] Cardano was born in Pavia, Lombardy, Italy, on September 24, 1501, as the illegitimate child of Fazio Cardano, a mathematically gifted jurist and a close personal friend of Leonardo da Vinci. Shortly before his birth, his mother, Chiara Micheria, had to move from Milan to Pavia to escape the Plague; her three other children died from the disease. In addition to his law practice, Fazio Cardano lectured on geometry at the University of Pavia and at the Piatti Foundation in Milan. Chiara lived apart from Fazio for many years, but, later in life, they married. Gerolamo was often sick and unhappy as a child. He received his education from his overbearing father and became his assistant. Fazio Cardano wanted his son to study law, but Gerolamo was more attracted to science and philosophy. His father's lessons on geometry had awakened his interest in the subject. After an argument with his father, he entered the University of Pavia in 1520 to study medicine. In 1524, the university had to close because of the Italian Wars (1521–1526), and Cardano moved to the University of Padua to complete his studies. He graduated with a doctorate in medicine in 1525. Cardano was a brilliant student but his eccentric and confrontational style did not earn him many friends. His father had died shortly after Gerolamo's move to Pavia, and the small bequest was soon eaten up. To improve his finances, Cardano had turned to gambling. His understanding of probability gave him an advantage over his opponents and, for some time, he indeed managed to make a living from playing card games, dice, and chess. After earning his doctorate, he repeatedly applied to join the College of Physicians in Milan, where his mother still lived. In spite of his excellent professional skills, he was not admitted, because of his reputation as a difficult person with strong opinions. The discovery of Cardano's illegitimate birth gave the college an official reason to reject his application. Without a membership in the College of Physicians, he could not practice as a physician in Milan, so he went to a small village near Padua and set up a small medical practice there. In 1531, Cardano married Lucia Bandarini, the daughter of a neighbor. Cardano's practice did not earn him enough money to support a family, and another attempt to get approved by the College of Physicians failed. To improve his finances, he started gambling again. But this time he lost more often than he won, and, after a series of losses, he even had to pawn his wife's jewelry to pay the bills. With no future perspectives in the countryside, Cardano and his wife moved to Milan, hoping to improve their situation.

After some struggling in the beginning, Cardano was fortunate to get his father's former position as a lecturer in mathematics at the Piatti Foundation. The job was not too demanding and so he had enough free time to treat a few patients in private. His successful treatments steadily increased his reputation as a medical doctor, and even members of the College of Physicians began to consult him for advice. Moreover, he gained wealth and influence through his appreciative upper-class patients. He was able to quit his teaching position, but he did not give up mathematics. In 1539, he finally received his admission from the College of Physicians, with the help of some influential supporters. However, in the same year, he published his first two mathematical books and he got in contact with Tartaglia, a self-taught mathematician who had achieved fame by winning a public competition on solving cubic equations at the University of Bologna.

Tartaglia was born as Niccolò Fontana in Brescia, Italy, in 1499.[3] His father was a dispatch rider who traveled to neighboring towns to deliver mail. Tragically, he was murdered by robbers, leaving his wife, the six-year-old Niccolò, and his two siblings in poverty. The poor family suffered further, when French troops invaded Brescia during the War of the League of Cambrai against Venice. When the French finally broke through, they massacred the inhabitants of the city. Niccolò and his family sought sanctuary in the local cathedral, but the French entered, and a soldier sliced Niccolò's jaw

Figure 11.2. Niccolò Fontana Tartaglia. Wikimedia Commons: https://commons.
wikimedia.org/wiki/File:Niccol%C3%B2_Tartaglia.jpg. Year of creation: 1572;
Copper engraving (?); Rijksmuseum; Artist unknown.

and palate with a saber and left him for dead. Niccolò was lucky to survive, but his speech was permanently impaired due to the injury, and so he got the nickname "Tartaglia" ("stammerer").

As an adult, he grew a beard to cover his scars. He worked as an engineer and bookkeeper in the Republic of Venice, and he was also a very ambitious, self-taught mathematician. At that time, Italian scholars had to prove their academic competence in public competition against other scholars, setting each other mathematical challenges. In 1535, Tartaglia won a famous public competition by demonstrating that he was able to solve a special type of cubic equations (equations with terms including x^3 as the highest power), something that had been considered impossible. He had discovered a method to solve equations of the form $x^3 + bx + c = 0$ as well as $x^3 + ax^2 + c = 0$, but not the general case of a cubic equation $x^3 + ax^2 + bx + c = 0$. Tartaglia's findings caught the attention of Cardano and, after having failed to find the solution by himself, he persistently tried to persuade Tartaglia to reveal his method. Tartaglia finally agreed to tell Cardano his solution, but Cardano had to promise that he would keep it secret. Tartaglia divulged his formula in the form of a poem, to make it more difficult to read for other mathematicians, in case the paper fell into the wrong hands:

When the cube and things together
Are equal to some discrete number,
Find two other numbers differing in this one.
Then you will keep this as a habit
That their product should always be equal
Exactly to the cube of a third of the things.
The remainder then as a general rule
Of their cube roots subtracted
Will be equal to your principal thing
In the second of these acts,
When the cube remains alone,
You will observe these other agreements:
You will at once divide the number into two parts
So that the one times the other produces clearly
The cube of the third of the things exactly.
Then of these two parts, as a habitual rule,
You will take the cube roots added together,
And this sum will be your thought.
The third of these calculations of ours
Is solved with the second if you take good care,
As in their nature they are almost matched.
These things I found, and not with sluggish steps,
In the year one thousand five hundred, four and thirty.
With foundations strong and sturdy
In the city girdled by the sea.[4]

This poem refers to one particular case of a cubic equation, $x^3 + bx + c = 0$ (in general, a cubic equation may also contain a term with x^2). With knowledge of Tartaglia's solution, Cardano soon succeeded in solving the general case of a cubic equation, $x^3 + ax^2 + bx + c = 0$, and only a short time later, Cardano's student and secretary, Ludovico Ferrari (1522–1565), devised a similar method to solve quartic equations (fourth-degree equations with terms of highest degree x^4). Of course, Cardano was eager to publish these results in his next book, with credit to Tartaglia for the decisive steps, but he had sworn that he would not reveal Tartaglia's method. However, in 1543, Cardano traveled to Bologna, where he was shown the notebooks of the deceased mathematician Scipione del Ferro (1465–1526). Cardano discovered that del Ferro had solved cubic equations long before Tartaglia (yet for less general cases) and thus felt no longer bound to keep the solution secret. In 1545, Cardano published his book *Artis Magnæ, Sive de Regulis Algebraicis Liber Unus* (*Book Number One about the Great Art, or The Rules of Algebra*; see fig. 11.3), or *Ars Magna*, as it is more commonly known, which is considered one of the greatest scientific treatises of the early Renaissance.

The book included the methods to solve cubic equations, and Cardano explained the history of their discovery as follows: "In our own days Scipione del Ferro of Bologna has solved the case of the cube and first power equal to a constant, a very elegant and admirable accomplishment. . . . In emulation of him, my friend Niccolò Tartaglia of Brescia, not wanting to be outdone, solved the same case when he got into a contest with his [Scipione's] pupil, Antonio Maria Fior, and, moved by many entreaties, gave it to me."[5] The solution to the general quartic equation was also contained in the book, with credit to Cardano's student, Ludovico Ferrari. In spite of the proper credits, Tartaglia felt betrayed by Cardano, who had broken his word by publishing Tartaglia's solution. In the following year, Tartaglia published a book in which he laid out his side of the story and personally attacked Cardano. The dispute lasted for many years, and it culminated in a public contest between Tartaglia and Ferrari in Milan. Tartaglia soon realized that Ferrari understood cubic and quartic equations better than he did, then left Milan before the contest was over. Thus, Ferrari won by default and Tartaglia's reputation diminished dramatically. As a result of the lost contest, he became effectively unemployable as a mathematician and had to resume his previous job in Venice, where he died in poverty. However, finding the solution to cubic equations was not Tartaglia's only contribution to mathematics. He is also remembered for the first translation of Euclid's *Elements* into a modern language (Italian), and he was the first to apply mathematics to the study of ballistic curves (paths of cannonballs).

Cardano's *Ars Magna* established him as one of the leading mathematicians of his time, and the solution methods for cubic and quartic equations are clearly the most important results in this work. Today, most mathematicians would acknowledge del Ferro, Tartaglia, and Cardano for the solution of cubic equations, since they all made decisive contributions. Let us

HIERONYMI CAR
DANI, PRÆSTANTISSIMI MATHE
MATICI, PHILOSOPHI, AC MEDICI,
ARTIS MAGNÆ,
SIVE DE REGVLIS ALGEBRAICIS,
Lib.unus. Qui & totius operis de Arithmetica, quod
OPVS PERFECTVM
inscripsit,est in ordine Decimus.

HAbes in hoc libro,studiose Lector,Regulas Algebraicas (Itali, de la Cos
sa uocant) nouis adinuentionibus,ac demonstrationibus ab Authore ita
locupletatas,ut pro pauculis antea uulgo tritis,iam septuaginta euaserint.Ne
que solum , ubi unus numerus alteri,aut duo uni,uerum etiam,ubi duo duobus,
aut tres uni equales fuerint,nodum explicant.   Hunc aut librum ideo seor-
sim edere placuit,ut hoc abstrusissimo, & plane inexhausto totius Arithmeti
cæ thesauro in lucem eruto , & quasi in theatro quodam omnibus ad spectan
dum exposito, Lectores incitaretur,ut reliquos Operis Perfecti libros, qui per
Tomos edentur,tanto auidius amplectantur,ac minore fastidio perdiscant.

Figure 11.3. The title page of the *Ars Magna* (*The Great Art*),
first published in 1545.

now reveal their method to obtain a solution, using modern mathematical symbols: We may consider a general cubic equation, $x^3 + ax^2 + bx + c = 0$, noting that any cubic equation can be written in this form (we divide an equation by any nonzero number to obtain an equivalent equation). Substituting $x = y - \dfrac{a}{3}$, we write the equation as

$$\left(y - \frac{a}{3}\right)^3 + a \cdot \left(y - \frac{a}{3}\right)^2 + b \cdot \left(y - \frac{a}{3}\right) + c = 0$$

and, by expanding the powers, we obtain:

$$y^3 - 3y^2 \cdot \frac{a}{3} + 3 \cdot y \left(\frac{a}{3}\right)^2 - \left(\frac{a}{3}\right)^3 + a\left(y^2 - 2y \cdot \frac{a}{3} + \left(\frac{a}{3}\right)^2\right) + b\left(y - \frac{a}{3}\right) + c = 0$$

which reduces to $y^3 = py + q$, where

$$p = \frac{a^2}{3} - b \text{ and } q = 4\left(\frac{a}{3}\right)^3 + \frac{b \cdot a}{3} - c$$

Then we let $y = z + w$. Expanding the left side of the equation, $(z + w)^3 = p\,(z + w) + q$, we obtain $z^3 + w^3 + 3zw\,(z + w) = p\,(z + w) + q$. Obviously, this equation is satisfied if $z^3 + w^3 = q$ and $3zw = p$. We use the second of these equations to write $w = \dfrac{p}{3z}$ and substitute this into the first equation, thereby arriving at

$$z^3 + \left(\frac{p}{3z}\right)^3 = q \,.$$

Multiplying this equation by z^3 we get

$$z^6 + \left(\frac{p}{3}\right)^3 = qz^3, \text{ which we can write as } \left(z^3\right)^2 - qz^3 + \left(\frac{p}{3}\right)^3 = 0,$$

a quadratic equation for z^3 with roots

$$\frac{q}{2} \pm \sqrt{\left(\frac{q}{2}\right)^2 - \left(\frac{p}{3}\right)^3} \,.$$

Since $z^3 + w^3 = q$, one root represents z^3, and the other one, w^3. Recalling that $y = z + w$, we finally get

$$y = \sqrt[3]{\frac{q}{2} + \sqrt{\left(\frac{q}{2}\right)^2 - \left(\frac{p}{3}\right)^3}} + \sqrt[3]{\frac{q}{2} - \sqrt{\left(\frac{q}{2}\right)^2 - \left(\frac{p}{3}\right)^3}}$$

which is indeed a solution of the cubic equation $y^3 = py + q$. This however, implies that we have also solved the original equation $x^3 + ax^2 + bx + c = 0$, as x can easily be inferred from y via the transformation $x = y - \dfrac{a}{3}$.

Cardano's wife, Lucia, died in 1546; they had three children: Giovanni, Chiara, and Aldo Battista. Cardano became a professor of medicine at the University of Padua and one of the most sought-after physicians in aristocratic circles. He later wrote that he even turned down offers from the kings of Denmark and France, and the queen of Scotland. Contrary to his great professional success, his private life was overshadowed by tragedies. Cardano's eldest son, Giovanni Battista, was executed for poisoning his wife; and his youngest son, Aldo Battista, became a gambler with friends of dubious character. Around 1563, Cardano wrote another important mathematical treatise, *Liber de Ludo Aleae* (*Book on Games of Chance*), which was not published until 1663. It contains the first systematic treatment of probability, based on the game of throwing dice as an example, as well as a section on effective cheating methods. He also made significant contributions to our knowledge of hypocycloids (curves traced out by a point on circles that roll inside another circle), which was published in *De proportionibus* in 1570. Cardano was an extremely prolific writer and one of the last polymaths of the Renaissance. He wrote more than 230 books in diverse fields, of which 138 were published; apart from mathematics and medicine, he was also interested in physics, biology, chemistry, astronomy, and even astrology. In fact, he was put in jail on the charge of heresy for casting the horoscope of Jesus Christ. After he was released, he went to Rome, and the pope not only forgave him but even granted him a pension. Cardano spent the last years of his life in Rome, where he died on September 21, 1576. He is reported to have correctly predicted the exact date of his own death, but it has been claimed that he achieved this by committing suicide.

~

CHAPTER 12

John Napier:
Scottish (1550–1617)

The use of the decimal point in mathematics and the application of logarithms to do calculations are largely due to the influence of Scottish mathematician John Napier, who was born on February 1, 1550, at Merchiston Castle in Edinburgh, Scotland. Napier was part of a noble family, so he was privately tutored until the age of thirteen, after which he entered St. Salvator's College at St Andrews. It is not known how long he stayed there, but it is believed that he traveled through Europe for several years, until 1571, when he returned to Scotland. In 1572, Napier married sixteen-year-old Elizabeth Stirling, also a product of nobility. He fathered two children in this marriage; but, unfortunately, his wife, Elizabeth, died in 1579. Shortly thereafter, Napier married Agnes Chisholm, with whom he had ten more children.

Later, in 1608, when his father died, Napier moved with his family into Merchiston Castle in Edinburgh, where he resided for the rest of his life. In 1614, Napier published a book, *Mirifici Logarithmorum Canonis Descriptio* (*A Description of the Wonderful Table of Logarithms*); it contained 147 pages, of which 90 were consumed with tables of numbers related to natural logarithms. This effort had begun around 1594, when he computed millions of entries, which took an incredibly long time. We can see from figure 12.2 what the first page of these listings looked like, which allows us to appreciate the intensive labor this must have required. Napier was asked to show the benefit of the logarithms system. He responded by showing that finding

Figure 12.1. John Napier. (Engraving by Samuel Freeman, 1835,
based on a 1616 painting in the University of Edinburgh and published
in Robert Chambers, ed., *A Biographical Dictionary of Eminent Scotsmen*, vol. 4
[Glasgow: Blackie & Son, 1835], facing page 88.)

the geometric mean can be done far more efficiently using logarithms than
simply doing the straight-out arithmetic—particularly when the numbers
are very large. Napier also realized that by using logarithms, calculations
that typically required multiplication or division, could now be reduced to
addition or subtraction of exponents—or, in this case, logarithms.

In addition, his book also treated the topic of spherical trigonometry.
However, his invention of logarithms first became popular when Henry
Briggs visited him in 1615, and helped him revise the logarithm tables. Essentially, Napier's work with mathematical computation had a great deal
of influence on the scientists of his time, including the famous Danish astronomer Tycho Brahe (1546–1601). Sadly, just as the book was gaining
popularity, Napier died, in Edinburgh on April 4, 1617.

Let us consider some of the advantages that Napier's work had provided during the seventeenth century and beyond. This was a time when

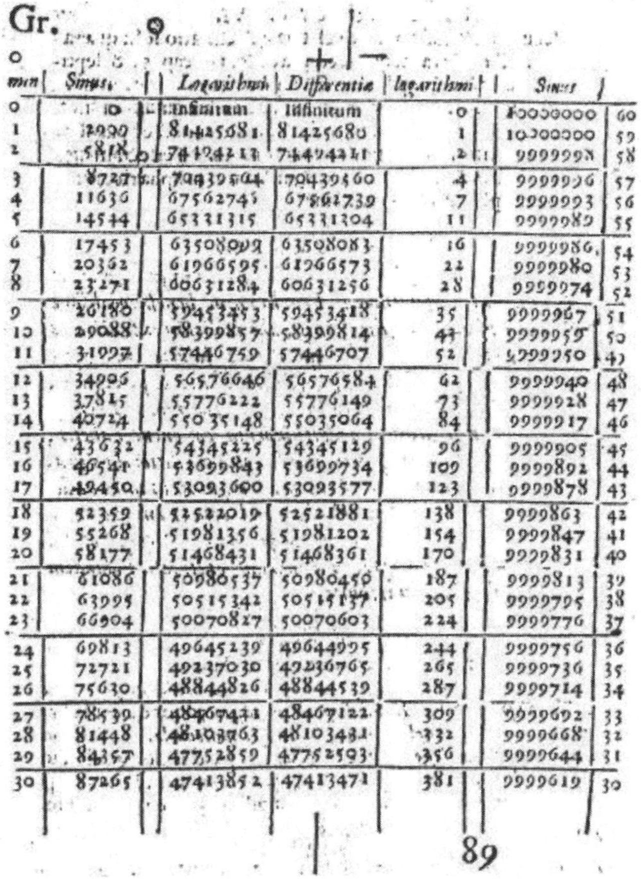

Gr. 9

min	Sinus		Logarithmi	Differentia	Logarithmi		Sinus	
0	.1 10	Infinitum	Infinitum	·0·	10000000	60
1	2909	81425681	81425680	1	10000000	59
2	5818	74124113	74494141	2	9999998	58
3	8727	70439564	70439560	4	9999996	57
4	11636	67562745	67862739	7	9999993	56
5	14544	65331315	65331304	11	9999989	55
6	17453	63508092	63508083	16	9999986	54
7	20362	61966595	61966573	22	9999980	53
8	23271	60631284	60631256	28	9999974	52
9	26180	59453453	59453418	35	9999967	51
10	29088	58399857	58399814	43	9999955	50
11	31997	57446759	57446707	52	9999950	49
12	34906	56576646	56576584	62	9999940	48
13	37815	55776222	55776149	73	9999928	47
14	40724	55035148	55035064	84	9999917	46
15	43632	54345225	54345129	96	9999905	45
16	46541	53699843	53699734	109	9999892	44
17	49450	53093690	53093577	123	9999878	43
18	52359	52522019	52521881	138	9999863	42
19	55268	51981356	51981202	154	9999847	41
20	58177	51468431	51468361	170	9999831	40
21	61086	50980537	50980450	187	9999813	39
22	63995	50515342	50515137	205	9999795	38
23	66904	50070827	50070603	224	9999776	37
24	69813	49645239	49644995	244	9999756	36
25	72721	49237030	49236765	265	9999736	35
26	75630	48844826	48844539	287	9999714	34
27	78539	48467411	48467122	309	9999692	33
28	81448	48103763	48103431	332	9999668	32
29	84357	47752859	47752593	356	9999644	31
30	87265	47413852	47413471	381	9999619	30

89

Figure 12.2. The first page of Napier's tables.
(Image from Landmarks of Science Series, NewsBank-Readex.)

mathematicians were very concerned about calculation, and how it could be simplified or mechanized. Toward that end, Napier developed a mechanical system, known as Napier's Rods, which we might consider a very early version of a calculator.[1] This system for performing multiplication, using only addition through the use of specially constructed strips, is shown in figure 12.3. The rods can be made out of cardboard, wood, or—what John Napier used when he invented this system of multiplication—bone, which provides us with another name for this method: Napier's Bones. Before reading what follows, you may want to spend a little time examining this figure to try to understand the logic of the construction.

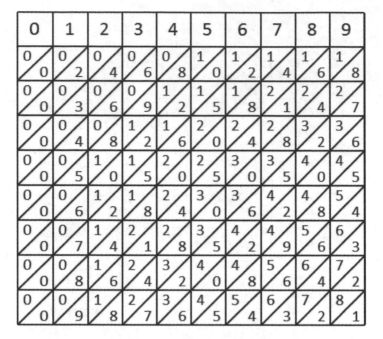

Figure 12.3.

There are ten vertical rods, each of which shows a specific column from the multiplication table written in a peculiar manner. Notice how the rod marked at the top with the digit 5 continues downward, with each of the multiples of 5 (10, 15, 20, etc.) written such that the tens digit is above the diagonal line and the ones digit is below the line. The same principle can be observed in the other rods: the fifth entry on the number 7 rod is 35, which is the same as the product $5 \cdot 7 = 35$. (Notice also that we put a 0 above the slash in entries where the product is less than 10.)

These rods can be rearranged freely, permitting us to construct the numbers we want to multiply and then to perform the computation using only addition. How is this possible? Let's look at an example to learn about the method Napier devised.

We will choose two numbers at random, in this case, 284 and 572, and then select the rods whose top digits will allow us to construct one of the numbers. It doesn't matter which of these two numbers we choose to represent first. Thus, in this example we will construct 572, selecting the rods numbered 2, 5, and 7, and then putting them in the correct order to match our number: 5, 7, 2 (see fig. 12.4).

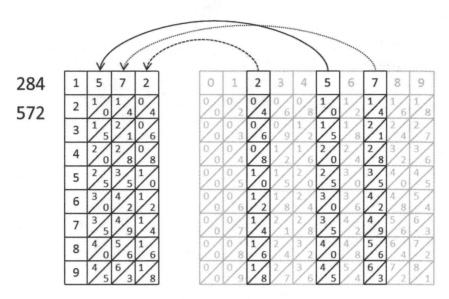

Figure 12.4.

We have written the digits 1 through 9 along the left-hand side, in a single column. In Napier's original construction, these numbers were written or engraved along the side of a shallow box inside of which the rods fit snugly. If you choose to re-create this example on your own, writing the numbers on a sheet of paper will work just fine, as long as you make sure to line up the tops of your rods appropriately as you place them.

As you may have already guessed, the next step is to identify the rows that we will need to construct our second number. With physical rods, it would not be possible to extract these rows, but for our illustration we will rearrange them, as indicated by the arrows, to form the number 284 (again maintaining proper alignment; see fig. 12.5).

In order to illustrate the next step, we will de-emphasize the boundaries between the rods, while highlighting the diagonal lines (see fig. 12.6). At the end of each diagonal, we have created a space where our sum can be written, as indicated by the dashed arrows. It looks like our product will be a six-digit number, since there are six diagonals in our final computation.

We find the sum of each diagonal, moving from the bottom-most diagonal to the upper-most diagonal; whenever that sum is greater than 9, we write the digit to be carried in a slightly smaller font inside the box, as well as at the head of the next diagonal, again, moving from the bottom-most

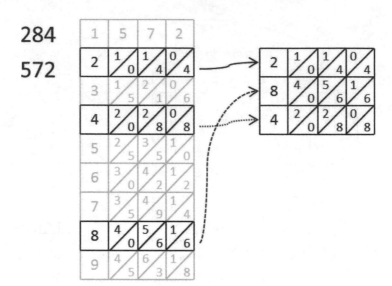

284

572

Figure 12.5.

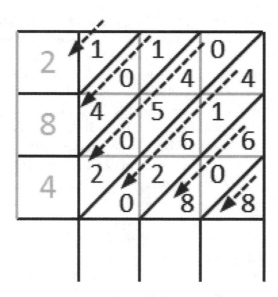

Figure 12.6.

to the top-most diagonal. Looking at the second diagonal, you can see the sum: 8 + 0 + 6 = 14, which means the tens digit of our final product will be 4, while the 1 is carried to the head of the third diagonal and added to the other numbers there, as shown in figure 12.7.

Proceeding along each diagonal, we see the sums are 8, 14, 14, 12, 6, and 1. Reading these in order from the top down and from left to right, without the carried digits, we get 1 6 2 4 4 8, which indicates that our final product is 162,448. You can check this with your calculator to verify that it is, indeed, correct!

How does this method work? Normally, the multiplication of two numbers is performed by successive digit multiplications and positional arithmetic. When you do multiplication according to the method typically taught in elementary school, you place one number above the other with a line underneath and multiply pairs of digits. As you do so, you write the ones digit of each product below the line, carrying the tens digits when

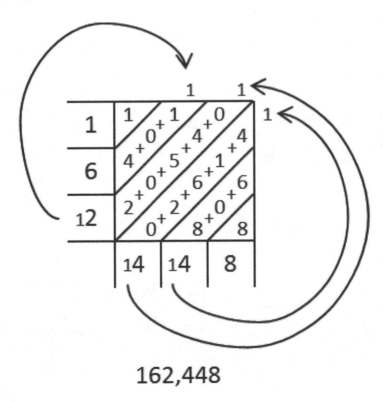

162,448

Figure 12.7.

necessary, and taking the sum of the partial products at the end of the process. To illustrate how Napier's Rods work, we will break down this process
step-by-step. Recall, our problem is 572 · 284.

The first step is to multiply 572 by 4.

The products of these multiplications are 4 · 2 = 8, 4 · 7 = 28, and
4 · 5 = 20. Carrying the 2 from the second multiplication and adding these
together, we get a partial total of 2288 (see the left side of fig. 12.8). Notice
that 2288 is the same result we would obtain from adding the diagonals of
row 4 of figure 12.4 (see the right side of fig. 12.8).

Repeating this process for the second digit, 8, we get 8 · 572 = 4,576,
which again is the same result we get from adding the terms in the diagonals of the eighth row of figure 12.4. According to the algorithm we know
from elementary school, we insert a 0 in the ones column, leaving us with
45,760 in the new, final row (see fig. 12.9).

Next, we multiply 2 · 572 and insert two 0s, giving us 114,400, the first
four digits of which we recognize from the second row of Napier's Rods (see
fig. 12.10).

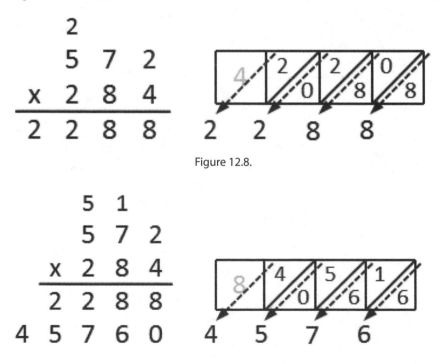

Figure 12.8.

Figure 12.9.

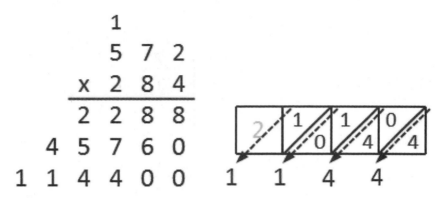

Figure 12.10

Finally, we add these three numbers (2,288 + 45,760 + 114,400). Again we find a result of 162,448, which is indeed the correct product, as shown in figure 12.11.

To complete our illustration of this method, we will do one final alteration: Instead of adding the digit products as we go, we will instead write the products as we did when constructing Napier's Rods, using a leading 0 for any number less than 10. Each product will be written with the appropriate offset, but in the same order we used when performing the previous operations.

Alongside this, we will draw the relevant portion of our Napier's Rods, this time rotated one-quarter turn, as we have in figure 12.12.

Do you notice anything interesting? That's right—each digit we produced using the traditional method of multiplication is also present in the

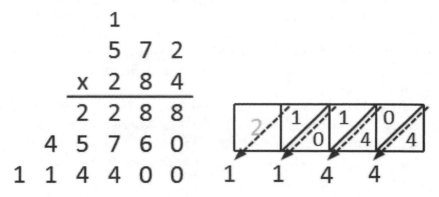

Figure 12.11

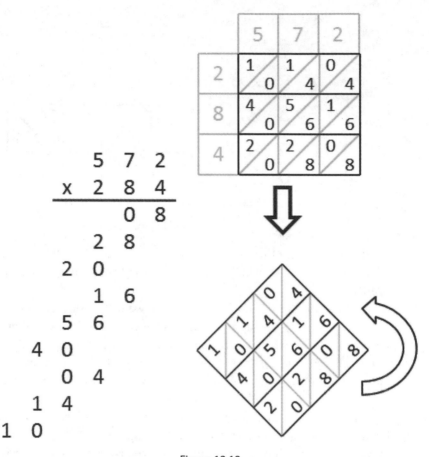

Figure 12.12

Napier's Rods representation, in the proper column! Also, if you look close-ly at the dark-outlined rows, you will notice that there is an exact corre-spondence between these rows and the respective digit products. So, for instance, the final three rows on the left (moving upward from the bottom) are 10, 14, and 04, and these same numbers are in the top column in the figure on the right.

As we have observed, the method of Napier's Rods is mechanically identical to our elementary-school algorithm, but it can make keeping track of the positions of each digit much easier. As an added advantage, it helps us to avoid multiplication errors—after all, most of us can do addition much more accurately than we can do multiplication!

Although by today's standards this method of calculation is rather primitive, one must bear in mind that for the times, when this was developed in the seventeenth century, it was seen as a great step forward. So much so, in fact, that Napier is probably better known for Napier's Rods, or Napier's Bones, than he is for his introduction of logarithms as an efficient method of calculation going forward centuries.

~

CHAPTER 13

Johannes Kepler:
German (1571–1630)

Two of the most significant aspects in the field of astronomy as it relates to our solar system are that the sun is the center of the solar system and that the planets revolve around the sun on an elliptical path. The former of these two brilliant discoveries was made by the Polish mathematician Nicolaus Copernicus; the second, by Johannes Kepler. Kepler was born in the city of Weil der Stadt (about twenty miles west of today's city of Stuttgart, Germany) on December 27, 1571, to a family that seemed to be faltering financially, even though his grandfather was the mayor of their town. Kepler's father left the family and died soon thereafter—all when Kepler was five years old. As a child, he impressed his neighbors with his amazing mathematical memory and facility. His early exposure to astronomical events such as the Great Comet of 1577 and a lunar eclipse in 1580 had an indelible effect on him for the rest of his life by generating an interest in the field.

From a case of smallpox as a child, Kepler's vision was impaired, and his hands' dexterity was limited throughout the rest of his life, which was in part a hindrance in his observational work in astronomy. Despite these physical limitations, he still rose to the top of the field through his phenomenal achievements. In 1589, he began his studies at the University of Tübingen. His initial studies focused on philosophy and theology, but his mathematical talents shifted his interests toward that field of study. He was fascinated by Copernicus's theory that the sun was the center of our universe—something not universally accepted in his day. In 1594, after

Figure 13.1. Johannes Kepler.
(Portrait painted in 1610 by an unknown artist.)

completing his university studies at age twenty-three, he accepted a position as a teacher of mathematics and astronomy at Grazer evangelische
Landschaftsschule im Paradeishof, the Protestant School in Graz, Austria.
While in Graz, in 1596, Kepler published a major work, *Mysterium Cosmographicum* (Cosmographic Mystery), which further supported Copernicus's belief of a heliocentric universe.

It must be said that, in part, Kepler's interest in having the sun at the
center of the universe was motivated by his firm theological convictions regarding God. This publication was his attempt to define the sizes of spheres
and the orbits in which they travel around the sun. Kepler used a rather
strange model to depict the planetary motion about the sun.

Referring to the model shown in figure 13.2, according to Kepler, the
outside sphere would represent the path of Saturn, then a sphere inscribed
in a cube, which is inscribed in the first sphere, would represent the path of
Jupiter. Then he inscribed a regular tetrahedron inside this smaller sphere,
which would contain another sphere representing the path of Mars. Inside
that sphere, a regular dodecahedron would be inscribed, and following

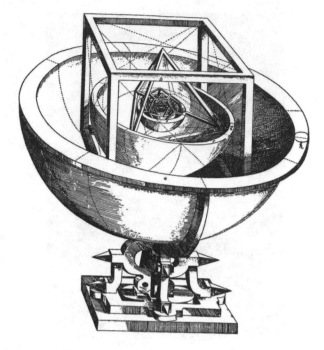

Figure 13.2. (Johannes Kepler, *Mysterium Cosmographicum*, 1596.)

this scheme, would separate Mars from Earth. Then a regular icosahedron would separate Earth from Venus, and, finally, a regular octahedron inscribed in the last sphere would separate Venus from Mercury. As confusing as this would seem to the modern eye, it gave Kepler some new fame. After his book was published in Tübingen, a copy was sent to Prague to the Danish astronomer Tycho Brahe, who was one of the foremost astronomers of his day, and who was in search for a mathematician to support his research. In 1600, Kepler met Brahe in the town of Benátky nad Jizerou, about twenty-two miles from Prague, where Brahe's observatory was being built. There, he spent time analyzing the data found, and, as time went on, he was given more access to the findings that were originally kept under guard. At first, it was a rocky relationship, but eventually they came to agree on salary and living arrangements.

Kepler took this position and then eventually succeeded Brahe when, shortly thereafter, Brahe died in 1601. The next eleven years were Kepler's most productive. Brahe assumed that the planets were traveling in circular orbits. Working with Brahe's conclusions, Kepler found that Mars was traveling in an elliptical orbit with the sun at one focus. To come up with this finding, known as Kepler's first law of planetary motion, he worked with numerous astronomical observations. These observations were then extended to the other planetary motions. Kepler's second law of planetary motion was to state that the line segment joining a planet to the sun in its elliptical path swept out equal areas in equal time periods. He published this in 1609 in a book titled *Astronomia Nova* (*New Astronomy*). The process to establish these two laws required many observations and calculations, which are still available to us today. In 1990, an American science historian, William H. Donahue, translated this book into English and found that Kepler had made some errors in his calculations; or, as we might say today, he fudged the data a bit in order to draw the conclusions for which he then later became very famous. Donahue says that this should not detract from Kepler's findings.[1] More than likely, this fudging compensated for the primitive tools Kepler was forced to use in the seventeenth century. This was reported in the *New York Times* on January 23, 1990.

Building upon Galileo's telescope, Kepler, who was already fascinated with optics, presented a new design for a telescope, using two convex lenses, where the final image is inverted. This was originally referred to as a Keplerian telescope and today is referred to as an astronomical telescope. He published the results of his work, in 1611, in *Dioptrice*.

Now, looking back to Kepler's personal life, we find that in 1597, Kepler married Barbara Müller, a widow from a wealthy background. Shortly after their marriage, they had two children, both of whom died at birth; the couple later had additional children. Then, in 1611, Kepler's seven-year-old son died, which upset him tremendously, and, making matters worse, shortly thereafter, his wife died. This was the time when, in Prague, tolerance for Protestants was not very good. At first, Kepler was given special dispensation to practice Lutheranism on his own, but eventually he decided rather than to convert to Catholicism, he would leave Prague with his remaining children and settle in the town of Linz, Austria. Now with children to care for and no wife, he was in search of someone to fill that role. During the next two years he considered eleven different matches. Then, in 1613, he married his second wife, the twenty-four-year-old Susanna Reuttinger, who took over the household, splendidly cared for his three children, and bore him another six children, of which only three survived.

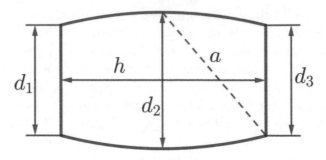

Figure 13.3.

At his wedding to Susanna, Kepler noticed that the volume of the wine barrels was measured by a rod that was slipped into the barrel diagonally (see fig. 13.3). This measurement amazed Kepler and he began to consider it more scientifically.

The volume of the barrel was then calculated as $V = 0.6\ a^3$, where a is the length of the rod inside the barrel. The same method was applied to barrels of different shapes, and Kepler recognized that the formula cannot give correct results in all cases, since the precise mathematical relation between the exact volume of the barrel and the length of a must depend on its proportions. This problem fascinated him, and he undertook a study of volumes, which could be seen as solids created through the revolution of a plane surface. First, let us consider the simplest case of a cylindrical barrel, for which $d_1 = d_2 = d_3$. The volume of a cylinder with diameter d and height h is $V = \pi \dfrac{d^2}{4} h$. By the Pythagorean theorem, we have $a^2 = \left(\dfrac{h}{2}\right)^2 + d^2$. A typical relation between h and d for the wine barrels Kepler studied would be $h = 2d$. Using this in the equation for a^2, we obtain $a = \sqrt{2}\ d$, or $d = \dfrac{a}{\sqrt{2}}$ and, moreover, $h = 2\dfrac{a}{\sqrt{2}} = \sqrt{2}\ a$. Expressing d and h in terms of a in the equation for the volume, we get $V = \dfrac{\sqrt{2} \cdot \pi}{8} a^3$ and $\dfrac{\sqrt{2} \cdot \pi}{8} = 0.55536\ldots$, explaining the approximation formula $V \approx 0.6a^3$. Kepler found a much more accurate formula by approximating the curvature of the barrel by a parabola. This formula is now generally known as Simpson's rule[2] (however, in Germany and Austria, it is also called Kepler's rule):

$$V = \frac{h \cdot \pi}{24} \cdot \left(\left(d_1\right)^2 + 4\left(d_2\right)^2 + \left(d_3\right)^2 \right)$$

To derive this formula, one has to calculate the area of a parabolic segment, a problem solved by Archimedes using methods essentially equivalent to the concept of integration in modern calculus. We will not provide a proof of Kepler's rule (or Simpson's rule), but present the result by considering a special approximation without using parabolas, from which Kepler's formula for the volume of a wine barrel can also be obtained. To this end, we divide the barrel into three rotational volumes of equal height: a frustum[3] of a cone at the bottom, a cylindrical part in the middle, and another frustum at the top (see fig. 13.4).

The volume of the cylinder is equal to the area of the base times the height, that is, $\frac{(d_2)^2 \pi}{4} \cdot \frac{h}{3} = \frac{h \cdot \pi}{12}(d_2)^2$. To get an approximate value for the volume of a frustum, we replace it by a cylinder, whose base area is equal to the mean cross-sectional area of the frustum. Note that this will give not the exact volume of a frustum but a fairly good approximation if the area at the bottom does not differ too much from the area at the top (i.e., if the frustum is close to a cylinder). The mean cross-sectional area of the first frustum is

$$\frac{1}{2} \cdot \left(\frac{\pi}{4} \cdot (d_1)^2 + \frac{\pi}{4} \cdot (d_2)^2 \right) = \frac{\pi}{8} \left((d_1)^2 + (d_2)^2 \right)$$

and the mean cross-sectional area of the second frustum is, therefore,

$$\frac{\pi}{8} \left((d_2)^2 + (d_3)^2 \right)$$

Multiplying these expressions by $\frac{h}{3}$ and adding the volume of the cylindrical middle part, we get a good approximation of the total volume of the wine barrel, if its shape does not deviate too much from a cylinder:

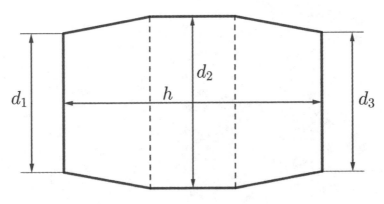

Figure 13.4.

$$V = \frac{h}{3} \cdot \left(\frac{\pi}{8} \left(\left(d_1\right)^2 + \left(d_2\right)^2 \right) + \frac{\pi}{8} \left(\left(d_2\right)^2 + \left(d_3\right)^2 \right) \right) + \frac{h \cdot \pi}{12} \left(d_2\right)^2 = \frac{h \cdot \pi}{24} \cdot \left(\left(d_1\right)^2 + 4\left(d_2\right)^2 + \left(d_3\right)^2 \right)$$

which is in exact agreement with the rule Kepler proposed. Kepler published his findings on generating volumes in 1615, in a work that was later refined by the Italian mathematician Bonaventura Cavalieri (1598–1674). Today we know this procedure as Cavalieri's principle. In 1619, while in Linz, Kepler published his second work on cosmology, titled *Harmonices mundi* Book V (*Harmony of the World*, Book V).

Of particular importance in this publication is what we today call Kepler's third law of planetary motion, which states that, for any two planets, the ratio of the squares of their periods (i.e., one complete revolution) is equal to the ratio of the cubes of the mean radii of their orbits. In 1621,

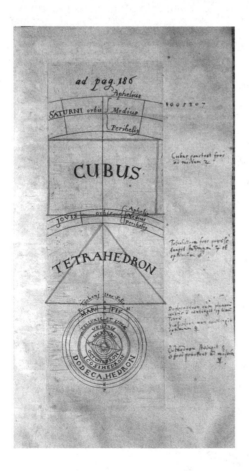

Figure 13.5. *Harmonices mundi*, Book V, 1619.

Kepler published a second version of his ideas in *Mysterium Cosmograph-icum*—although in much shorter form and correcting much of what he consequently found to be inaccurate in the original work. Kepler remained very active in computational matters, not the least of which was motivated by the writings of Scottish mathematician John Napier, on logarithms. He published his computations in *Rudolphine Tables* in 1627, which included eight-place logarithms.

His book on a variety of tables is often considered one of his greatest works, but, as we can see in figure 13.6, it was not published until 1627. This is because there were legal disagreements with Tycho Brahe's heirs as to ownership of the work. In 1625, the religious tension in Europe between Catholics and Protestants led to the Counter-Reformation, placing most of Kepler's library under seal. By 1626, the city of Linz was besieged, and Kepler was forced to move to the city of Ulm, Germany, where the *Rudolphine Tables* was ultimately printed. The battles subsided in 1628, and Kepler became an official advisor to the Bohemian general Albrecht von Wallenstein (1583–1634); in this capacity, he provided astronomical calculations for various astrologers. He did quite a bit of traveling in his last years, between Prague, Linz, and Ulm. He fell ill in the city of Regensburg, Germany, and died on November 15, 1630. He was buried there, but, in later years, the cemetery was decimated; therefore, there is no gravesite for Kepler, but there are monuments to him in both Prague and Linz.

Let us now review the three laws of planetary motion that have bestowed upon Kepler the most fame.

Kepler's First Law: Planets move about the sun in an elliptical orbit, with the sun at one focus of the ellipse (see fig. 13.7).

Kepler's Second Law: The speed of a planet traveling along an elliptical orbit, with the sun at one focus, is such that the line joining the sun and the planet sweeps out equal areas during equal time periods. (See fig. 13.8, where the two shaded regions have equal areas.)

Kepler's Third Law: The square of the orbital period of a planet is proportional to the cube of the semi-major axis of its orbit, as shown in figure 13.9.

It must be said that these three laws that Kepler discovered are truly amazing, especially given the limitations of the astronomical tools that were available during his time. As mentioned earlier, we can consider him extraordinarily clever to have seen these relationships despite the inaccuracy of his measurements.

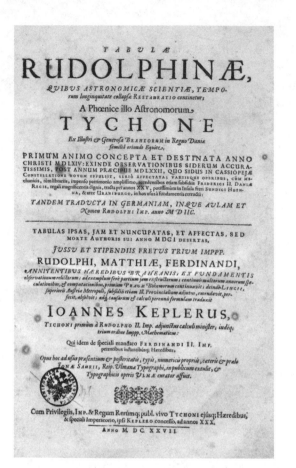

Figure 13.6. *Rudolphine Tables*, 1627.

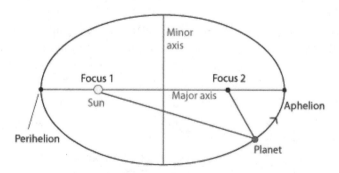

An elliptical orbit of a planet
(greatly exaggerated)

Figure 13.7. (Image by Brian Ventrudo, *The One Minute Astronomer* [Mintaka, 2008], ebook.)

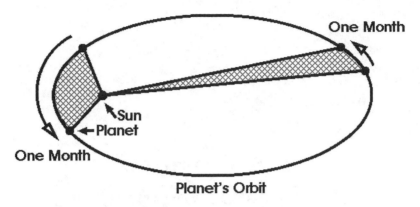

Figure 13.8. (Image by Brian Ventrudo, *The One Minute Astronomer*
[Mintaka, 2008], ebook.)

$$\frac{P^2}{p^2} = \frac{a^3}{a^3}$$

1st Planet

P = Orbital Period
a = Semi-Major Axis

2nd Planet

p = Orbital Period
a = Semi-Major Axis

Figure 13.9.

Chapter 14

René Descartes:
French (1596–1650)

The 2017 Nobel Prize in Physiology or Medicine was awarded to Jeffrey C. Hall, Michael Rosbash, and Michael W. Young, "for their discoveries of molecular mechanisms controlling the circadian rhythm."[1] The circadian rhythm is a general term for biological processes oscillating with a period of approximately twenty-four hours, meaning that these processes are adapted to the rotational period of the earth. Life on Earth has developed biological clocks, which are responsible for regulating metabolism, hormone levels, sleep, and other aspects of our physiology. Our inner clock is independent of sunlight and would maintain our periods of sleepiness and wakefulness in a cycle of about twenty-four hours, even if we were living in complete darkness or underground, without any natural light sources. A mismatch between our inner clock and our environment or lifestyle has adverse effects on our well-being and productivity. For example, when we travel across several time zones, we may experience jet lag, a temporal phenomenon caused by a large shift between the time kept by our internal biological clock and the external time dictated by the environment. However, not all people living in the same time zone have synchronized inner clocks; there is some individual variation, depending on lifestyle, work schedule, and biological factors. For instance, our circadian rhythm gets shifted during adolescence, letting us go to bed late. This explains why most teenagers are "night owls." Yet, for biological reasons, teenagers also need more hours of sleep than adults do. If the school day starts early, it therefore becomes

Figure 14.1. René Descartes.
(Portrait by Franz Hals, oil on canvas, ca. 1649–1700.)

difficult for teenagers to get enough sleep. Besides, getting up out of bed is quite a challenge when your inner clock commands you to sleep for another two or three hours.

When the French philosopher and mathematician René Descartes (1596–1650) was about ten years old, he was sent as a boarding student to the Jesuit College at La Flèche, established in 1604 by King Henri IV of France. There, he was granted an unusual privilege, which would probably be a dream come true for any teenager regularly struggling with early-morning wake-up calls. While the other boys at boarding school had to get up at five o'clock in the morning, René was officially allowed to stay in bed until eleven o'clock because of his weak physical condition and his frequent health problems. Young René enjoyed sleeping late, but even after he woke up, he often remained in bed for several hours. Alone and without any distraction, he would meditate on the knowledge and subjects he was taught at La Flèche, including classical studies, traditional Aristotelian philosophy, science, and mathematics. He later described his education at La Flèche as follows:

I had been assured I could acquire a clear and certain knowledge of all
that is useful in life. I had an extreme desire to learn. But as soon as I
had completed the course of study, at the end of which one is usually
received into the rank of the learned, I entirely changed my opinion.
For, I found myself embarrassed by so many doubts and errors, that I
thought I had gained nothing else from trying to instruct myself, than
to have more and more discovered my ignorance.[2]

There was only one subject in school that was free of doubt, a quality
Descartes found very appealing:

I took pleasure, above all, in mathematics, because of the certainty
and the absoluteness of its reasons; but I had not yet found out its true
use; and, thinking that it served only for the mechanical arts, I was
astonished that, its foundations being so firm and solid, nothing had
ever been built on them that was more exalted.[3]

Throughout his life, Descartes kept the habit of spending a considerable
amount of daytime in bed, pondering fundamental philosophical and scientific questions. Isolated from the world around him, he was able to focus
his mind and reach a state of deep contemplation, disputing with himself
about what we know and what we can know. Today, he is most famous for
the Latin phrase "*cogito ergo sum.*" Translated into English, this is, "I think,
therefore I am." It means that we cannot doubt of our existence while we
doubt; that is, the very act of thinking serves as a proof of the reality of one's
own mind. Descartes admired the strict deductive reasoning used in mathematics and the absolute certainty of mathematical results. He thought that
all science and philosophy should be based on mathematics. This is meant
in the sense that we cannot accept anything as certain unless it can be deduced by a complete and rigorous chain of evidence from already-secured
knowledge or observations of nature and scientific experiments. In one of
his most important publications, the *Discours de la méthode pour bien con-
duire sa raison et chercher la vérité dans les sciences* (*Discourse on the Method
of Rightly Conducting One's Reason and of Seeking Truth in the Sciences*),
Descartes writes:

The long chains of simple and easy reasoning by means of which geometers are accustomed to reach the conclusions of their most difficult

demonstrations led me to imagine that all things, to the knowledge of which man is competent, are mutually connected in the same way, and that there is nothing so far removed from us as to be beyond our reach, or so hidden that we cannot discover it, provided only, we abstain from accepting the false for the true, and always preserve in our thoughts the order necessary for the deduction of one truth from another.[4]

However, in order to deduce truths from other truths, one first needs a starting point or a premise on which to base further reasoning. This basis must be provided by statements that are taken to be true or are accepted without controversy or question. Statements of this type are called postulates or axioms; they cannot be deduced from more-elementary statements. The foundations of modern mathematics are based on minimal lists of axioms. For instance, many facts in arithmetic can be derived from more-basic facts; however, as one traces these basic facts back to even more basic facts and continues with this process, one will eventually end up at statements that cannot be reduced any further. It turns out that number theory (the study of integers) can be built upon the so-called Peano axioms, named after Giuseppe Peano (1858–1932) (see chap. 39). These axioms consist of five statements formulated in the language of mathematical logic and defining the set of natural numbers in terms of properties that are independent of their concrete representation. The Peano axioms can be viewed as the first principles of number theory. Descartes's famous *"cogito ergo sum"* plays a similar role for modern Western philosophy. In his *Principia Philosophiae* (*Principles of Philosophy*), he characterizes first principles as follows:

> First, they must be so clear and so evident that the human mind cannot doubt of their truth when it attentively considers them; and second, the knowledge of other things must depend upon these Principles in such a way that they may be known without the other things, but not vice versa.[5]

However, Descartes not only tried to put philosophy on a mathematical foundation but also made important contributions to mathematics itself, making him one of the most influential mathematicians of his time. We will reveal some of his achievements in mathematics and some of the mathematical notations and concepts attributed to him, while giving a brief overview of his life.

René Descartes was born at his grandmother's home in the commune La Haye en Touraine, France (renamed La Haye-Descartes in 1802 and renamed again to Descartes in 1967), on March 31, 1596. His father, Joachim, had studied law and was a counselor at the court of justice. When René was only one year old, his mother, Jeanne, died in childbirth; René was sent back to this maternal grandmother, who would care for him. His father remarried in 1600, and René continued living with his grandmother, together with two older siblings. After his education at La Flèche, he entered the University of Poitiers, where he studied law, following the paths of his father and his maternal uncle, René Brochard, who was a deputy and judge at the Estates-General in Poitiers. Descartes received his degree and legal license in 1616. Nothing was known about the content of his thesis until 1981, when a curator for the Sainte-Croix Museum (Poitiers) made an unexpected discovery while reframing a seventeenth-century engraving that had been hanging in a museum restaurant. He found, stuffed in the back of the engraving, a public broadsheet that had been printed in 1616 and announced the oral thesis defense of René Descartes (see fig. 14.2).

The broadsheet contains an affectionate dedication to his uncle René Brochard and a list of forty statements summarizing his thesis. However, Descartes did not pursue a career as a lawyer or a judge any further. Although the theory of law has some similarities to mathematics, since deductive reasoning is used to draw conclusions from legal text, there is also a profound difference: mathematical statements are universal; legal text is invented by humans and has nothing to do with nature. Descartes decided to stop devoting his time and energy to the study of books from which he would not learn anything about nature. In his *Discourse on the Method*, he recalls:

> I entirely abandoned the study of letters, resolving to seek no knowledge other than that which could be found in myself or else in the great book of the world.[6]

In 1618, Descartes entered a military school in Breda, Netherlands, where he studied mathematics and physics, to become a military engineer. After serving in the armies of Maurice of Nassau and Maximilian of Bavaria, he spent a lot of time between 1620 and 1628 traveling through northern and southern Europe and developing his ideas for a philosophy based on the concept of mathematical proofs. In Paris, he was in regular contact

Figure 14.2. Broadsheet advertising Descartes's oral thesis defense in 1616.

with the French priest and mathematician Marin Mersenne (1588–1648), who had also studied at La Flèche and who encouraged him to publish his thoughts on philosophy and science. Although Paris was one of the intellectual centers of the world at that time, Descartes left Paris in 1628 and returned to the Netherlands, seeking a secluded place without distraction, to work on his ideas for a new philosophy. In 1633, he completed his first major treatise on physics, *The World* (French: *Traité du monde et de la lumière*), based on the heliocentric model of the sun and the planets developed by Polish mathematician Nicolaus Copernicus. But when Descartes learned that Galileo Galilei was condemned by the Catholic Church for defending Copernicanism in his *Dialogue*, he shied away from publishing his treatise. However, fragments of this work appeared together with his famous *Discourse on the Method* in 1637, in three essays: "Les Météores" ("The Meteors"), "La Dioptrique" ("Dioptrics"), and "La Géométrie" ("Geometry"). The *Discourse on the Method* consists of six parts and is one of the most influential works of modern philosophy; it is also the first important modern philosophical work that was not written in Latin. Descartes wrote it in French to make his work accessible to everyone, not only to scholars. He describes a universal method of deductive reasoning, applicable to all sciences. In the second part, Descartes presents the four basic rules of his method, revealing his inspiration from mathematical proofs:

1.   Accept nothing as true that is not self-evident.
2.   Divide problems into their simplest parts.
3.   Solve problems by proceeding from simple to complex.
4.   Recheck the reasoning.

Of the three essays supplementing the *Discourse on the Method*, "La Géométrie" is by far the most important. With this work, Descartes revolutionized mathematics by bringing together Euclidean geometry and algebra to what is now known as analytic geometry. Before Descartes, geometry and algebra had essentially been separate fields, whereby geometry was considered to be fundamental. Descartes discovered that with the help of a coordinate system defined by two directed perpendicular lines, geometric shapes can be described by algebraic equations. As the simplest examples, figure 14.3 shows two straight lines and a circle centered at the origin, together with their corresponding equations relating the *x*-coordinate and the *y*-coordinate.

Coordinate systems with perpendicular axes are called Cartesian coordinates in honor of Descartes (*Cartesius* is the Latin version of "Descartes"). By representing geometric objects such as straight lines, circles, and other curves by algebraic equations, suddenly, problems in geometry could be solved by algebraic methods, and vice versa. In particular, geometric relations between geometric objects, such as tangency, intersection points, and so on, could be described by corresponding algebraic equations. Today, this is also referred to as analytic geometry. For example, using his method of coordinates, Descartes discovered the following algebraic solution to a special case of Apollonius's problem of finding a circle touching three mutually tangent circles, as shown in figure 14.4.

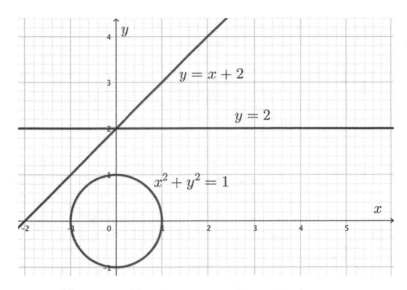

Figure 14.3. Curves can be represented by algebraic equations.

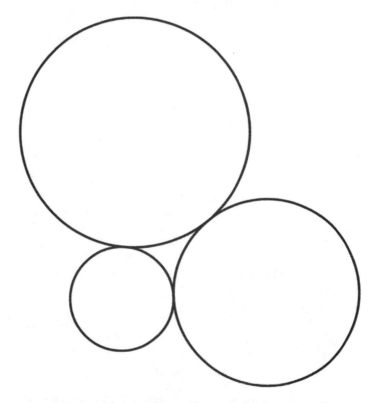

Figure 14.4. Three circles tangent to each other externally.

If the three given circles have radii r_1, r_2, and r_3, then the radius of the fourth circle is determined by the equation

$$\frac{1}{r_4} = \frac{1}{r_1} + \frac{1}{r_2} + \frac{1}{r_3} \pm 2\sqrt{\frac{1}{r_1 r_2} + \frac{1}{r_2 r_3} + \frac{1}{r_3 r_1}}$$

where the ± sign reflects the fact that there are two solutions to the problem—shown as the two dashed circles in figure 14.5. This statement is known as Descartes's theorem.

The introduction of Cartesian coordinates was a milestone in the history of mathematics and also a fundamental ingredient in the development of calculus by Isaac Newton and Gottfried Wilhelm Leibniz.

The convention of using the letters x, y, and z to represent variables and letters a, b, c, . . . for known quantities is also attributed to Descartes, as is the use of superscripts for powers or exponents, such as x^2.

After his seminal *Discourse on the Method*, Descartes continued to produce important works concerning both mathematics and philosophy, the most comprehensive of which is the *Principia Philosophiae*, published in Amsterdam in 1644. By 1649, Descartes had become one of the most famous philosophers and scientists in Europe, in spite of not holding any

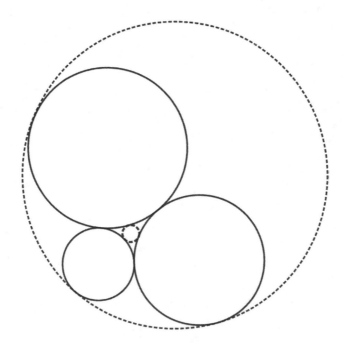

Figure 14.5. The dashed circles are tangent to all three solid circles.

academic position. Yet he had always preferred to be left alone and to work isolated from the world, without distractions. He kept his residence in the Netherlands a secret and held contact with the scientific community only through exchanging letters with Marin Mersenne, who was one of the very few persons who knew Descartes's address.

In 1649, Queen Christina of Sweden invited Descartes to her court in Stockholm, to organize a scientific academy and to teach her. Some persuasion was necessary until Descartes finally accepted and moved to Sweden in the middle of the winter. He had kept the habit of lying in bed until eleven o'clock throughout his lifetime, but the twenty-two-year-old Christina of Sweden insisted upon receiving philosophy lessons at five o'clock in the morning. The fifty-three-year-old Descartes now had to break the rhythm he was accustomed to and fight against his inner clock, which weakened him and made him more susceptible to infections. Walking to the queen's palace every morning in the cold Swedish winter did the rest, and he soon caught a cold from which he developed pneumonia. Only ten days after falling ill, René Descartes died on February 11, 1650, in Stockholm.

CHAPTER 15

~

Pierre de Fermat:
French (1607–1665)

Today, research in mathematics is fragmented into a vast number of branches and sub-branches, all having their own defined community of mathematicians. In pure mathematics alone, there are more than one hundred journals of excellent reputation; they are devoted to particular fields and publish only high-quality articles containing important new findings. Mathematicians meet at international conferences to present their work, plan collaborations, and exchange ideas. Unless a conference is exclusively devoted to a very special topic, most of the presentations will be comprehensible only to a small minority of the audience. This is simply a consequence of the high degree of specialization to which mathematicians must adhere in order to get a chance to make new and relevant contributions to their field of research. In fact, even for the most outstanding mathematicians, it has become virtually impossible to be an expert in several different branches at the same time. Needless to say, it is totally inconceivable that someone who is not a professional mathematician would be able to obtain significant new insights anywhere near the forefront of current research. In the seventeenth century, mathematics had not yet branched into so many almost-disjointed subjects, and, at least in principle, it was still possible for "spare-time mathematicians" to gain enough knowledge and competence to not only correspond with and earn the respect of renowned mathematicians in academic positions but also even pursue their own ideas with success—and, in some cases, eventually contribute pioneering work. This is

Figure 15.1. A seventeenth-century portrait of Pierre de Fermat.

certainly true for the French mathematician Pierre de Fermat (1607–1665), who is still famous for his work in mathematics, even though he actually was a lawyer who could occupy himself with mathematics only alongside his professional obligations at the Parliament[1] of Toulouse in France.

Fermat was born in the fall of 1607, in Beaumont-de-Lomagne, France, where his father, Dominique Fermat, a wealthy leather merchant, served three one-year terms as one of the four consuls governing the town. Pierre's mother, Claire de Long, was a noblewoman. She died in childbed in 1615. There is little evidence regarding Pierre's school education, and we don't know whether he had a mentor in mathematics at school or what motivated his interest in mathematics. What *is* known is that he studied at the University of Orléans, where he received a bachelor's degree in civil law in 1626. He then moved to Bordeaux to work as lawyer. In Bordeaux, Fermat got in contact with the lineographer and mathematician Jean de Beaugrand, and he also formed a lifelong friendship with Etienne d'Espagnet, who had inherited a huge and very well-equipped library from his father, the Renaissance polymath Jean d'Espagnet. The mathematics section of the library contained works by Euclid, Apollonius of Perga, and François Viètes (1540–1603), also known as Franciscus Vieta, who was, in fact, a friend of Jean d'Espagnet. Fermat eagerly read these books, thoroughly studying the presented material and adding his own notes in the margins (see fig. 15.2).

Having gained extensive knowledge in mathematics, Fermat began his own mathematical investigations, concerned with tangents to algebraic curves and finding minima and maxima of functions. In parallel, he reconstructed Apollonius's lost work *De Locis Planis*, described in some detail by Pappus of Alexandria. It contained propositions relating to loci that are either straight lines or circles. Fermat inherited a fortune when his father died in 1628. He bought the office of a deceased councilor at the parliament

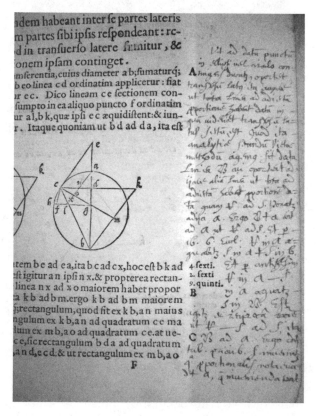

Figure 15.2. Handwritten notes of Pierre Fermat in a
Latin transcription of Apollonius's *Conics*.

in Toulouse and became his successor there. With his inauguration as a government official in 1631, he was entitled to change his name from Pierre Fermat to Pierre de Fermat, a right he himself, however, never made use of. In the same year, he married Louise de Long, his fourth-degree cousin. They had eight children, five of whom survived into adulthood. In 1636, Fermat's friend Pierre de Carcavi went to Paris and met Marin Mersenne, whom he told about Fermat's mathematical research. Mersenne then wrote to Fermat, and they began a correspondence that lasted until Mersenne's death in 1648. Fermat stayed in Toulouse for the rest of his life, visiting only nearby towns from time to time. He never traveled any further than to Bordeaux,[2] so his communication with other mathematicians was restricted to writing letters. With Mersenne as a mediator, Fermat corresponded with Galileo Galilei, Blaise Pascal, John Wallis, Christiaan Huygens, and René Descartes. Through his correspondence, Fermat quickly became known as a leading mathematician, although he rarely published his results, because he did not want to spend too much time on polishing the proofs for publication. Yet the letters that have survived, in conjunction with posthumously published writings and notes, clearly show that he made pioneering works in several fields of mathematics. Independent of Descartes, Fermat developed analytic geometry. There was a famous controversy between Fermat and Descartes regarding the soundness of their mathematical methods, with Descartes finally giving in and writing the following to Fermat:

> . . . seeing the last method that you use for finding tangents to curved lines, I can reply to it in no other way than to say that it is very good and that, if you had explained it in this manner at the outset, I would have not contradicted it at all.[3]

Fermat's method of finding the tangent to a curve was based on calculating the differential of the function describing the curve, essentially in the same way that differential quotients are computed in an elementary calculus course today. However, Fermat was lacking the concept of a limit to justify his calculations. A mathematically consistent formulation of modern calculus was established thirty years later by Newton and Leibniz. Interestingly, Newton wrote that the ideas that led to his invention of calculus were inspired by "Fermat's way of drawing tangents."[4] Moreover, Fermat is recognized as a key figure in the historical development of the fundamental principle of least action in physics. The principle of least action is a generalization of Fermat's principle of least time, named Fermat's principle in his honor. It states that light travels between two given points along the path of shortest time. Fermat was able to deduce Snell's law of refraction from his principle of least time.

Surprisingly, Fermat was not really interested in physics, but, when reading Descartes's treatise on optics, "La Dioptrique," he discovered that Descartes's heuristic derivation of the law of refraction was based on circular reasoning. Descartes became angry about Fermat's critique of his work, and this was the beginning of their disputes. Fermat's mathematical correspondence was interrupted between 1644 and 1653; perhaps his duties at the parliament did not allow him to continue with his mathematical research during this period of time. However, in 1654, Fermat received a letter from Blaise Pascal, who wanted to discuss his calculations of probabilities. Their resulting correspondence laid the foundation of probability theory (see chap. 16). Yet, Fermat's main mathematical interest, if not obsession, was number theory. Unfortunately, none of the mathematicians he was in contact with shared his enthusiasm for this topic, as it was not considered very important at that time. He tried to persuade Pascal, as well as Huygens, to join him in his research in number theory, but he wasn't successful. Fermat, indeed, made some important contributions to number theory, but he was not interested in publishing his work. Concerning his discussions with Pascal on the calculation of probabilities, Fermat wrote to Carcavi:

> I am delighted to have had opinions conforming to those of M Pascal, for I have infinite esteem for his genius. . . . The two of you may undertake that publication, of which I consent to your being the masters, you may clarify or supplement whatever seems too concise and relieve me of a burden that my duties prevent me from taking on.[5]

Fermat enjoyed posing problems to the leading mathematicians of his time. However, he rarely provided complete proofs for his theorems; often, he only sketched the method, and it was left to others to fill in the gaps. Number theory was not very fashionable at this time, and more than one hundred years went by until Leonhard Euler took up Fermat's studies and gave full proofs to some of the results or conjectures that Fermat had formulated without providing rigorous proofs. In 1659, Fermat wrote a letter to Carcavi, intended for Huygens, in which he gave a brief summary of his accomplishments in number theory and also revealed some of his methods, in particular the method of infinite descent.[6] In the last paragraph, he wrote,

And Perhaps posterity will thank me for having shown it that the ancients did not know everything, and this account will pass into the mind of those who come after me as a "passing of the torch" to the next generation.[7]

In 1653, Fermat was struck down by the plague and survived, but this episode probably had some long-term effects on his health. In a 1660 letter to Pascal, who lived in Clermont-Ferrand, about 236 miles from Toulouse, Fermat suggested they meet halfway between the two towns, since "my health is not any better than yours."[8] In 1664, he felt that his life would soon come to an end and wrote his last will and testament. He kept working as a judge at the parliament as long as he could, and he died in Castres, France, at the age of fifty-seven, on January 12, 1665—just one week after his last official act.

The most famous theorem of Fermat, for which he is remembered today, has a fascinating history and is in many ways characteristic of both his style as a mathematician and his work's significance for later developments in mathematics. It is known as Fermat's last theorem, and it can be stated as follows:

The equation $x^n + y^n = z^n$ has no positive integer solutions for x, y, and z when $n > 2$.

For $n = 2$, we would obtain the Pythagorean theorem, $x^2 + y^2 = z^2$, and this equation has, in fact, infinitely many integer solutions, which are called Pythagorean triples. For instance, the numbers 3, 4, and 5 form a Pythagorean triple, since $3^2 + 4^2 = 5^2$. After having found one Pythagorean triple, one can easily generate infinitely many others by multiplying each of the three numbers by the same positive integer (e.g., multiplication by 2 yields the numbers 6, 8, 10, which also form a Pythagorean triple). Fermat's last theorem was first discovered by his son, Samuel, in the margin in his father's copy of an edition of Diophantus's *Arithmetica*, together with the note:

I have a truly marvelous demonstration of this proposition which this margin is too narrow to contain.[9]

Fermat's proof was never found; however, Samuel republished Diophantus's *Arithmetica*, along with his father's marginal notes, in 1670 (see fig. 15.3) This popularized Fermat's last theorem, which became a famous

Arithmeticorum Liber II. 61

interuallum numerorum 2. minor autem 1 N. atque ideo maior 1 N. + 2. Oportet itaque 4 N. + 4. triplos esse ad 2. & adhuc superaddere 10. Ter igitur 2. adscitis vnitatibus 10. æquatur 4 N. + 4. & fit 1 N. 3. Erit ergo minor 3. maior 5. & satisfaciunt quæstioni.

ς² ἐός. ὁ ἄρα μείζων ἔςαι ς² ἑνὸς μ̄² β̄. δήσει ἄρα ἀειθμὲς δ̄ μονάδας δ̄ τριπλασίονας ἔη μ̄² β̄. ἐ ἔτι ὑπερέχειν μ̄² ῑ. τρὶς ἄρα μετάδες ϛ μ̄⸖ μ̄² ῑ. ἴσαι εἰσὶ ςςιν δ̄ μονάσι δ̄. ᾗ ϑίνεται ὁ ἀειθμὸς μ̄² γ̄. ἔςαι ὁ μὲν ἐλάσσων μ̄² γ̄. ὁ δὲ μείζων μ̄² ε̄. ᾗ ποιῶσι τὸ πρόβλημα.

IN QVÆSTIONEM VII.

CONDITIONIS appositæ eadem ratio est quæ & appositæ præcedenti quæstioni, nil enim aliud requirit quàm vt quadratus interualli numerorum sit minor interuallo quadratorum, & Canones iidem hic etiam locum habebunt, vt manifestum est.

QVÆSTIO VIII.

PROPOSITVM quadratum diuidere in duos quadratos. Imperatum sit vt 16. diuidatur in duos quadratos. Ponatur primus 1 Q. Oportet igitur 16 — 1 Q. æquales esse quadrato. Fingo quadratum à numeris quotquot libuerit, cum defectu tot vnitatum quod continet latus ipsius 16. esto a 2 N. — 4. ipse igitur quadratus erit 4 Q. + 16. — 16 N. hæc æquabuntur vnitatibus 16 — 1 Q. Communis adiiciatur vtrimque defectus, & a similibus auferantur similia, fient 5 Q. æquales 16 N. & fit 1 N. ⁴⁄₅ Erit igitur alter quadratorum ²⁵⁶⁄₂₅. alter verò ¹⁴⁴⁄₂₅ & vtriusque summa est ⁴⁰⁰⁄₂₅ seu 16. & vterque quadratus est.

TON ἐπιταχθέντα τετράγωνον διελεῖν εἰς δύο τετραγώνους. ἐπιτετάχθω δὴ τ̄ ιϛ διελεῖν εἰς δύο τετραγώνους. καὶ τετάχθω ὁ πρῶτος δυνάμεως μιᾶς. δήσει ἄρα μονάδας ιϛ λείψει δυνάμεως μιᾶς ἴσας ἔη τετραγώνῳ. πλάσσω τ̄ τετράγωνον ἀπὸ ςς. ὅσων δὴ ποτε λείψει ποσῶν μ̄² ὅσων ἐςὶν ἡ τ̄ ιϛ μ̄² πλευρά. ἔςω ςς β̄ λείψει μ̄² δ̄. αὐτὸς ἄρα ὁ τετράγωνος ἔςαι δυνάμεων δ̄ μ̄² ιϛ λείψει ςς ιϛ. ταῦτα ἴσα μονάσι ιϛ λείψει δυνάμεως μιᾶς. κοινὴ προσκείσθω ἡ λεῖψις, ἐ ἀπὸ ὁμοίων ὅμοια. δυνάμεις ἄρα ε̄ ἴσαι ἀειθμοῖς ιϛ. ἐ ϑίνεται ὁ ἀειθμὸς ιϛ. πέμπτων. ἔςαι ὁ εἶς ος⁵⁶ εἰκοσοπέμπτων. ὁ δὲ μεθ εἰκοσοπέμπτων, ἐ οἱ δύο συντεθέντες ποιῶσι

ν εἰκοσοπέμπτα, ἤτοι μονάδας ιϛ. καὶ ἔςιν ἑκάτερος τετράγωνθ.

OBSERVATIO DOMINI PETRI DE FERMAT.

CVbum autem in duos cubos, aut quadratoquadratum in duos quadratoquadratos & generaliter nullam in infinitum vltra quadratum potestatem in duos eiusdem nominis fas est diuidere cuius rei demonstrationem mirabilem sane detexi. Hanc marginis exiguitas non caperet.

QVÆSTIO IX.

RVRSVS oporteat quadratum 16 diuidere in duos quadratos. Ponatur rursus primi latus 1 N. alterius verò quotcunque numerorum cum defectu tot vnitatum, quot constat latus diuidendi. Esto itaque 2 N. — 4. erunt quadrati, hic quidem 1 Q. ille verò 4 Q. + 16. — 16 N. Cæterum volo vtrumque simul æquari vnitatibus 16. Igitur 5 Q. + 16. — 16 N. æquatur vnitatibus 16. & fit 1 N. ⁴⁄₅ erit

ΕΣΤΩ δὴ πάλιν τὸν ιϛ τετράγωνον διελεῖν εἰς δύο τετραγώνους. τετάχθω πάλιν ἡ τ̄ πρώτου πλευρὰ ς² ἑνὸς, ἡ ἡ τ̄ ἑτέρας ςς ὁσωνδήποτε λείψει μ̄² ὅσων ἐςὶ ἡ τ̄ διαρμένου πλευρά. ἔςω δὴ ςς β̄ λείψει μ̄² δ̄. ἔσονται οἱ τετράγωνοι ὃς μὲν δυνάμεως μιᾶς, ὃς δὲ δυνάμεων δ̄ μ̄² ιϛ λείψει ςς ιϛ. βούλομαι τὰς δύο λοιπὸν συντεθείσας ἴσους ἔη μ̄² ιϛ. δυνάμεις ἄρα ε̄ μ̄² ιϛ λείψει ςς ιϛ ἴσαι μ̄² ιϛ. καὶ ϑίνεται ὁ ἀειθμὸς ιϛ πέμπτων.

H iiij

open problem of mathematics, attracting the attention of many great mathematicians.

In spite of countless attempts to prove it or to find a counter-example, the theorem remained a conjecture until 1994, when the British mathematician Andrew Wiles (1953–) finally succeeded, after working secretly on the problem for six years. The proof comprises more than one hundred pages and was published as the entire issue of May 1995 of the *Annals of Mathematics*, 358 years after Fermat's conjecture. The proof relies on very special techniques from modern mathematical theories developed in the twentieth century, and it is incomprehensible for mathematicians who are not working in closely related fields. For solving this famous problem, Wiles was awarded a multitude of prizes, including one of the most prestigious award for mathematicians, the Abel Prize. It is now believed that Fermat's proclaimed "proof" was highly questionable, although it cannot be completely ruled out that, indeed, he did have a truly remarkable proof. In any case, the unsuccessful attempts to prove his theorem—extending over a period of more than three hundred years—led to an astounding number of more important mathematical discoveries and theories, with fruitful applications in branches of mathematics seemingly not at all related to number theory. Although no leading mathematician really shared his interest in number theory during his lifetime, Pierre de Fermat has managed to engage generations of mathematicians in his preferred field of research for more than three centuries after his death; this alone is quite unique in the field of mathematics.

CHAPTER 16

~

Blaise Pascal:
French (1623–1662)

One of the greatest mathematicians of all time was Blaise Pascal, who was born in Clermont, Auvergne, France, on June 19, 1623. His father, Étienne Pascal, was a politician and a man of culture and intellectual distinction. Blaise Pascal's mother died when he was four years old; thus, he was reared by his father, along with his two sisters. Encouraged by his father, his early years found him deeply engaged in religious thinking. This often distracted him from other intellectual endeavors. When Pascal was seven years old, Étienne moved to Paris with his three children. This was about the time when he was heavily involved in teaching his children at home. Pascal was not physically well conditioned, yet this was compensated for with an exceptionally brilliant mind. Étienne was impressed at how quickly his young son would pick up new ideas of what was then considered the classical education. He kept mathematics at a distance from him, so as not to put too much strain on the young child. Frankly, this built up Pascal's curiosity about mathematics even more. Once the father realized his son's incredible mathematics talents, he gave him a copy of Euclid's *Elements*, perhaps one of the first compilations of geometry and other aspects of mathematics in a logical development. One of Pascal's sisters claimed that her younger brother had discovered Euclid's first thirty-two propositions in the same order that Euclid did, without referring to the book. It was the thirty-second proposition, that the sum of the angles of a triangle is equal to the sum of two right angles, that further demonstrated Pascal's unique talent.

Figure 16.1. Blaise Pascal. (Lithograph after E. Edelinck after F. Quesnel Jr.)

By the age of fourteen, Pascal was admitted to the weekly meetings of a group that eventually developed into the French Academy of Science. About the time when he was sixteen years old, and having been motivated by the work of the French mathematician Girard Desargues (1591–1661), he got involved in geometry and proved some of the most beautiful theorems in the field, one of which today bears his name (Pascal's theorem) and is very easy to demonstrate. All you need is a circle and a ruler. We demonstrate this theorem in figure 16.2, where we randomly select six points on a circle, then join these points consecutively, forming a hexagon inscribed in the circle. We then extend the three pairs of opposite sides (*AB* and *ED*; *BC* and *FE*; and *CD* and *AF*) so that they would intersect. (Of course, when you select your six points, avoid placing them such that you would have any pair of opposite sides parallel.) When we mark these points of intersection, *L*, *N*, and *M*, we find that these points will always lie on a straight line. Amazingly, this holds true for any six points on a circle (avoiding parallels, as mentioned above), and, curiously enough, it can also be extended to any six points on an ellipse. At first, other famous mathematicians of the times, such as René Descartes, refused to believe that such a discovery could be made by a sixteen-year-old boy. But, in time, it was properly accepted.

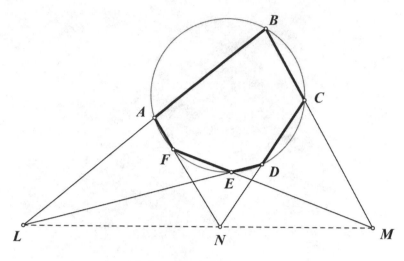

Figure 16.2.

Pascal had to pay a price for his brilliance. From the age of seventeen until the end of his life at age thirty-nine, he lived in physical pain, with sleepless nights and unpleasant days. Yet, he kept on working. At age nineteen, Pascal invented the first calculator machine (see fig. 16.3), in order to assist his father's computational work as a tax collector for the city of Rouen. This calculator was able to do addition and subtraction and was referred to as Pascal's calculator or *Pascaline*. There are currently four versions of this machine exhibited in the Musée des Arts et Métiers in Paris. At the time of its development, the machine was considered a luxury item and this motivated Pascal to continue to improve its functioning over the next 10 years.

The society in which he lived was tormented by religious upheaval, which to some extent affected Pascal as well, since his dear sister Jacqueline, who had supported him, entered a monastery in Pert-Royal. At age twenty-three, he suffered a temporary paralysis, but his intellectuality continued unabated. He continued to lead a rather turbulent life tortured by his family's involvement in various religious followings. In 1654, at age thirty-one, Pascal engaged in probably the most important contribution he had made to mathematics. That is, he embarked on a mathematical correspondence with Pierre de Fermat, which eventually became the basis for the theory of probability. During the year 1654, Pascal and Fermat challenged each other with mathematical problems that began to generate the development of the future field of probability, as we know it today. One of the early problems that was posed involved a game in which two players would gain points,

Figure 16.3. Pascal's calculator. (Wikimedia Creative Commons, photo by Rama, licensed under CC BY-SA 3.0 FR.)

with a specified number of points to win the game. The question was as follows: If the game is stopped before the end, how should the money be divided between the two players, considering the number of points each player has at the time of stop of the game? Here is a translation of one of these correspondences, from Fermat to Pascal in 1654:

> Monsieur. If I undertake to make a point with a single die in eight throws, and if we agree after the money is put at stake, that I shall not cast the first throw, it is necessary by my theory that I take $\frac{1}{6}$ of the total sum to be impartial because of the aforesaid first throw. And if we agree after that I shall not play the second throw, I should, for my share, take the sixth of the remainder, that is, $\frac{5}{36}$ of the total. If, after that, we agree that I shall not play the third throw, I should, to recoup myself, take $\frac{1}{6}$ of the remainder, which is $\frac{25}{216}$ of the total. And if, subsequently, we agree again that I shall not cast the fourth throw, I should take $\frac{1}{6}$ of the remainder, or $\frac{125}{1296}$ of the total, and I agree with you that that is the value of the fourth throw, supposing that one has already made the preceding plays. But you proposed in the last example in your letter (I quote your very terms) that if I undertake to find the six in eight throws, and, if I have thrown three times without getting it, and if my opponent proposes that I should not play the fourth time, and if he wishes me to be justly treated, it is proper that I

have $\frac{125}{1296}$ of the entire sum of our wagers. This, however, is not
true by my theory. For in this case, the three first throws, having
gained nothing for the player who holds the die, the total sum
thus remaining at stake, he who holds the die and who agrees to
not play his fourth throw should take $\frac{1}{6}$ as his reward. And if he
has played four throws without finding the desired point, and,
if they agree that he shall not play the fifth time, he will, nevertheless, have $\frac{1}{6}$ of the total for his share. Since the whole sum
stays in play it not only follows from the theory, but it is indeed
common sense that each throw should be of equal value. I urge
you therefore (to write me) that I may know whether we agree in
the theory, as I believe (we do), or whether we differ only in its
application. I am, most heartily, etc.

Fermat

This written exchange led to the beginning of questioning the likelihood of certain occurrences involving cards, point flips, and the like. This is
the very basic aspect of the field of probability.

Then there was Antoine Gombaud, the Chevalier de Méré (1607–
1684), who was a French gambler whose claim to fame was correspondence
with Pascal seeking help to understand why he was continuously losing at
a game of dice. This motivated Pascal to correspond further with Pierre de
Fermat, which generated a new field of mathematics and led to what we
know today as probability theory.

Let's take a look at what this exchange of ideas entailed. De Méré
was involved with two games of dice. The first game involves making a
bet with even odds on getting at least one six on four successive rolls of
the die. He knew that the likelihood of getting a six on one roll was $\frac{1}{6}$.
He then figured that on four rolls of the die, the probability would be
$\frac{4}{6} = \frac{2}{3}$. This, of course, was incorrect. This didn't stop him from betting,
and yet he seemed rather successful.

He extended his thinking by betting with even odds on getting at least
a double six on 24 rolls of a pair of dice. Figuring, correctly, the chance of
getting a double six on one roll of the pair of dice is $\frac{1}{36}$, once again he mis-
takenly assumed that getting the double six on 24 rolls of the pair of dice
would be $\frac{24}{36} = \frac{2}{3}$.

Since he began to lose a lot of money, he decided to seek help from his brilliant friend, Pascal. This further strengthened the correspondence between Pascal and Fermat, which ultimately led to a solution to the problem.

Let's take a look at the two games and see why the first game was profitable and the second game was not. Clearly, we know that when we roll the die there are six possible ways that it can land, which allows us to conclude that the probability of getting a six is $\frac{1}{6}$, and the probability of not getting a six is $\frac{5}{6}$. Therefore, considering de Méré's first game, we calculate that the probability of getting no six in four rolls of the die is

$$\frac{5}{6} \cdot \frac{5}{6} \cdot \frac{5}{6} \cdot \frac{5}{6} = \left(\frac{5}{6}\right)^4 = 0.4822531\ldots$$

It follows that the probability of getting at least one six on these four rolls is $1 - 0.4822531\ldots = 0.5177469\ldots$. We can interpret this for 100 games with approximately 52 successful rolls. Were he to play 1,000 games, he would win an average of 518 games. Winning more than half the games gave him an edge.

Now considering the second game, we recall that there were 36 possible outcomes when tossing two dice, of which only one was a double six. This gives us a probability of getting the double six as $\frac{1}{36}$; the probability of not getting a double six is $1 - \frac{1}{36} = \frac{35}{36}$. Therefore, the probability of not getting a double six on 24 rolls of the pair of dice is $\left(\frac{35}{36}\right)^{24} = 0.5085961\ldots$. As before, we can conclude that the probability of getting at least one double six on the 24 rolls of the pair of dice is $1 - 0.5085961\ldots = 0.4914039\ldots$. This indicates that de Méré would win only approximately 49 out of 100 games, which gives his opponent an edge, winning 51 out of 100 games. Problems of this sort were solved in the exchange between Pascal and Fermat, which led to what we know today as probability theory.

Throughout this time, Pascal made considerable use of the triangular arrangement of numbers that also bears his name today. In figure 16.4, we see this arrangement, where, beginning at the top, we have a 1, followed by a second row of two 1s; then, each succeeding row begins and ends with a 1, with each other number between the 1s being the sum of the two numbers diagonally above it on either side. This pattern then continues downward. Today, this arrangement of numbers is known as the Pascal triangle. Many number arrangements can be found on the Pascal triangle. For example, the sum of the numbers in each row is a power of 2, as shown in the right margin of figure 16.4.

When we look at figure 16.5, we also notice that, considered as numbers, we have a representation of powers of 11. This triangular arrangement of numbers is very helpful, to this day, when working with probability.

There are probably countless patterns that can be found on the Pascal triangle; however, one that surprises us most is the appearance of the Fibonacci numbers, which can be seen in figure 16.6.

It could be said that Pascal's name in today's recollection of the history of mathematics is being a co-inventor of the theory of probability, which seems to become increasingly more important in our everyday lives, from weather prediction to work in finance. On August 19, 1662, tortured with physical maladies and unpleasant mental conditions, Pascal's life came to an end in Paris, when he suffered convulsions and died at age thirty-nine.

	2^0	2^1	2^2	2^3	2^4	2^5	2^6	2^7	2^8	2^9	2^{10}
	1										
	1	1									
	1	2	1								
	1	3	3	1							
	1	4	6	4	1						
	1	5	10	10	5	1					
	1	6	15	20	15	6	1				
	1	7	21	35	35	21	7	1			
	1	8	28	56	70	56	28	8	1		
	1	9	36	84	126	126	84	36	9	1	
	1	10	45	120	210	252	210	120	45	10	1

Figure 16.4.

$$
\begin{array}{c}
11^{0}: \quad 1 \\
11^{1}: \quad 1 \quad 1 \\
11^{2}: \quad 1 \quad 2 \quad 1 \\
11^{3}: \quad 1 \quad 3 \quad 3 \quad 1 \\
11^{4}: \quad 1 \quad 4 \quad 6 \quad 4 \quad 1 \\
11^{5}: \quad 1 \quad 5 \quad 10 \quad 10 \quad 5 \quad 1 \\
11^{6}: \quad 1 \quad 6 \quad 15 \quad 20 \quad 15 \quad 6 \quad 1 \\
11^{7}: \quad 1 \quad 7 \quad 21 \quad 35 \quad 35 \quad 21 \quad 7 \quad 1 \\
11^{8}: \quad 1 \quad 8 \quad 28 \quad 56 \quad 70 \quad 56 \quad 28 \quad 8 \quad 1 \\
11^{9}: \quad 1 \quad 9 \quad 36 \quad 84 \quad 126 \quad 126 \quad 84 \quad 36 \quad 9 \quad 1 \\
11^{10}: \quad 1 \quad 10 \quad 45 \quad 120 \quad 210 \quad 252 \quad 210 \quad 120 \quad 45 \quad 10 \quad 1 \\
11^{11} \\
\end{array}
$$

Figure 16.5.

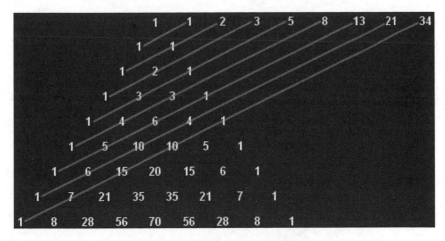

Figure 16.6.

CHAPTER 17

Isaac Newton: English (1642–1727)

As we embark on the life of perhaps the most important mathematician and physicist in history, we begin with an overview of his very modest lifestyle. Isaac Newton was born December 25, 1642, in a small town of Woolsthorpe-by-Colsterworth, in the County of Lincolnshire, England. Unfortunately, his father died three months before Isaac was born. As a tiny, premature baby, Newton was not expected to live, yet he did so for eighty-four years! When Newton was two years old, his mother remarried and decided to live with her new husband, the wealthy minister Barnabas Smith, whom the young Newton did not like. Newton was then left in the care of his maternal grandmother, Margery Ayscough. He seemed to be sour at his mother for marrying Smith. She then had three additional children in this second marriage.

Newton attended the King's School in Grantham, England, from the ages of twelve to seventeen; in addition to learning Latin and Greek, he had his first exposure to mathematics there. He returned to his original home in 1653 to live with his mother, who by then was widowed for a second time. There, in Woolsthorpe-by-Colsterworth, his mother urged him to do farming, which certainly did not suit Newton. Soon thereafter, the head of the King's School urged his mother to return him to school to finish his education, which she did, allowing Newton to begin to exhibit his brilliance and to become the school's top student. It should be said, though, that these early years caused him some psychological difficulties that accompanied him the rest of his life.

His excellence in school as well as the recommendation of his uncle, who was an alumnus of Trinity College at Cambridge, enabled him to be admitted to the college in 1661. Soon thereafter, he eventually earned a full scholarship. During his undergraduate studies, he began to immerse himself in Aristotle's work and philosophy. However, he also discovered the writings of French mathematician René Descartes, which guided Newton in a new direction and seemed to define the rest of his life. More specifically, Descartes's "La Géométrie" allowed him to focus on seeking algebraic solutions to geometric problems, which he felt was a much more conclusive way to solve problems. The works of Galileo and Kepler further strengthened his thinking in the direction of a heliocentric system of the universe. He made notes, titled *Quaestiones quaedam philosophicae*, in which he listed his thoughts on mechanical philosophy, guided by the best thinking of the times and his own imagination. In 1665, Newton expanded the binomial theorem (see fig. 17.4) by including fractional powers, which led him on the path to his development of what we today know as infinitesimal calculus.

In August 1665, Newton received his bachelor's degree and left the university, as it was closed for two years due to the Great Plague that dominated England. During the next two years, he stayed home, concentrating on private studies and further developing his theories of calculus, optics, and the law of gravitation.

Newton was elected as a fellow of Trinity College at Cambridge University in 1667, where he refused to become an ordained priest, which was previously a requirement for fellows at Cambridge. Initially, this precondition was not strictly enforced; but, by 1675, it became a requirement. Only through special permission from Charles II was Newton able to avoid becoming an ordained priest. In 1669, one year after receiving his master's degree, Newton succeeded Isaac Barrow, becoming the second Lucasian professor. During this time, Newton summarized his work by writing *De analysi per aequationes numero terminorum infinitas* (On Analysis by Infinite Series), which was shared with a limited audience and enabled his name to become better known. Soon thereafter, he wrote a revised form, *Tractatus de methodis serierum et fluxionum* (Treatise on the Methods of Series and Fluxions). In this work, he introduced the word *fluxions*, which is an indication of the birth of calculus. (More about this later.) In 1672,

Figure 17.1. Isaac Newton. (Portrait by Sir Godfrey Kneller, 1689.)

with his reputation for brilliance expanding, he was elected as fellow of the Royal Society.

It is also well known that about the same time as Newton's work on calculus became popular, the German mathematician Gottfried Wilhelm Leibniz (1646–1716) developed differential and infinitesimal calculus using completely different symbols. Leibniz's symbols are largely still in use today, as opposed to Newton's symbols, which are no longer used. It should be noted that Newton and Leibniz were in bitter disagreement regarding who should be credited with the development of calculus; this disagreement grew to the point where, beginning in 1699, members of the Royal Society started to accuse Leibniz of plagiarism. There is evidence to show that Newton generated this ill feeling, which continued until Leibniz's death in 1716.

Newton was probably best known for his discoveries in the field of physics, where he made great advances in the field of optics; in particular, he discovered the spectrum of light by splitting white light through a prism. He was also well known for his significant improvements in the development of a telescope. Yet he is probably best known for his three laws of motion, which were presented in his famous book *Philosophiæ Naturalis Principia Mathematica* (*Mathematical Principles of Natural Philosophy*), commonly known as the *Principia*, which was first published in 1687 (see fig. 17.2).

The first law of motion states that every object will remain at rest, or in uniform motion along a straight line, unless affected by an external force. The second law, sometimes referred to as the law of force and acceleration, states that a force on an object to accelerate is equal to the product of mass and acceleration, or, to put it another way, the acceleration of an object is directly proportional to the force and inversely proportional to the mass of the object. The third law states that for every action there is a reaction, that is, an action in the opposite direction.

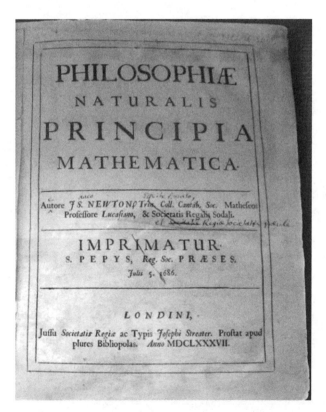

Figure 17.2. Newton's personal copy of *Principia Mathematica*, with his own comments included, in 1686.

There aren't many examples of Newton's mathematical discoveries that can be presented to the general readership, but we will offer just a few of them here. Together with the English mathematician Joseph Raphson (1648–1715), Newton developed what is today referred to as the Newton-Raphson method of square-root extraction. In 1690, Raphson published *Analysis Aequationum Universalis*, which included a method that is a simpler version of what Newton published in his *Method of Fluxions*. Newton wrote this latter work in 1671, but it was not published in English until 1736 (see fig. 17.3). Raphson was a strong supporter of Newton's work, especially when it came to crediting him, not Leibniz, with discovering calculus.

Let us now investigate how the Newton-Raphson method allows us to extract the square root of a number in a rather simple way, one that truly makes sense when compared to other more automatic algorithms that are

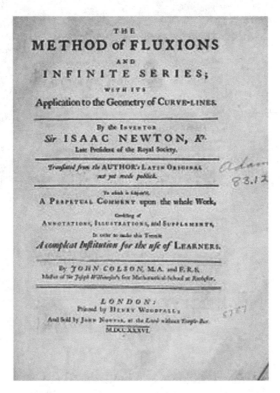

Figure 17.3. Newton's *Method of Fluxions*.

not as easily intuitively justifiable. Perhaps it easier to view this method through an example.

Consider finding $\sqrt{40}$. We know that this number is somewhere between $\sqrt{36} = 6$ and $\sqrt{49} = 7$. We will guess that the value is about 6.3. If this number were correct, and $\sqrt{40}$ were equal to 6.3, then $\frac{40}{6.3}$ would have to be equal to 6.3. But it is not the case, since $\frac{40}{6.3} \approx 6.35$.

Therefore, we know that the number we seek as the square root of 40 must be somewhere between 6.3 and 6.35. We will take the average of these two numbers: $\frac{6.3 + 6.35}{2} \approx 6.325$.

We now continue the process with $\frac{40}{6.325} \approx 6.3241$ and then we try to see if this value is actually the square root of 40. If it is the correct value, then by dividing 40 by 6.3241, the quotient would also have to be 6.3241. However, $\frac{40}{6.3241} \approx 6.32501$.

Once again, taking the average gets us another decimal place closer to the actual value of $\sqrt{40}$. Thus, $\frac{6.3241 + 6.32501}{2} \approx 6.324555$.

We can continue this process to get as accurate a value for the square root of 40 as we wish. Notice that with each step in the process we move one decimal place closer to the value of $\sqrt{40}$. When we compare this to the calculator-generated value for $\sqrt{40}$, we get 6.324555320336759 . . . , which allows us to see how nicely the Newton-Raphson method brings us to a fine approximation for the square root of 40.

We also credit Isaac Newton with the further development of the binomial theorem, which is presented to most folks during their high-school mathematics instruction. You may recognize the pattern by inspecting the first several applications shown in figure 17.4. Furthermore, if you focus on the coefficients of each of the terms in the binomial expansion, you will notice the Pascal triangle emerging.

In general form, the binomial theorem can be expressed in the following fashion:

$$\left(a+b\right)^n = a^n + \binom{n}{1}a^{n-1}b + \binom{n}{2}a^{n-2}b^2 + \binom{n}{3}a^{n-3}b^3 + \ldots + b^n$$

where $\binom{n}{r} = {}_nC_r = \dfrac{n!}{r!\left(n-r\right)!}$

However, Newton's contribution was highlighted by the fact that he was able to conclude that the binomial theorem was also true for fractional or irrational powers, where n could also be a fraction or even a value of the form of $\sqrt{k}$. Here, what was a finite sum now becomes an infinite series.

$$(a + b)^0 =$$ 1
$$(a + b)^1 =$$ $1a + 1b$
$$(a + b)^2 =$$ $1a^2 + 2ab + 1b^2$
$$(a + b)^3 =$$ $1a^3 + 3a^2b + 3ab^2 + 1b^3$
$$(a + b)^4 =$$ $1a^4 + 4a^3b + 6a^2b^2 + 4ab^3 + 1b^4$
$$(a + b)^5 =$$ $1a^5 + 5a^4b + 10a^3b^2 + 10a^2b^3 + 5ab^4 + 1b^5$

Figure 17.4. Binomial theorem.

Newton did so much in mathematics and physics that our short chapter could hardly touch on even a small fraction of his works. He was the first to employ coordinate geometry to solve Diophantine equations (which we encountered in chap. 8). Newton said that he preferred the geometrical methods to algebraic ones, as he felt that they were clearer and more rigorous. Further pursuits of these extensions take us to a more advanced level, which is beyond the scope of this book.

It is curious that Newton's psychological difficulties kept much of his work on pure mathematics shared with only his colleagues and other select correspondents, until 1704, when he published his book *Opticks*. At that time, he published works on the quadrature of curves and also on the classification of cubic curves. This was actually the first time that Newton published his ideas on the method of fluxions, the precursor of today's calculus. Although he hinted at it in his *Principia*, it was actually Leibniz's paper in 1684 that would put calculus in the public forum. As we mentioned earlier, there were many bilateral accusations of plagiarism, even though it is clear that both mathematicians came upon their discoveries independently. It should also be said that this experience once again demonstrated Newton's psychological imbalance. Newton's further mathematical publications evolve from his Cambridge lectures, which he delivered from 1673 to 1683, and which were first published as late as 1707.

By the 1690s, Newton began to delve into religious thinking, and he wrote about his interpretations of the Bible both literally and symbolically. There has been much controversy about Newton's belief of the doctrine of the Trinity in the New Testament, but there was no conclusive evidence to close the case. Despite these doubts, he was a devout Protestant and opposed any Catholic infiltration. In his various scientific writings later in life, Newton expressed a strong sense of God's providential role in nature.

In the years 1689–1690 and 1701–1702, Newton represented Cambridge University in the English Parliament. He was delighted to assume the post of warden of the Royal Mint in 1696; in this position, he was in charge of a major re-coining process in England. Three years later, he became the master of the Mint, a position that he held for the last thirty years of his life. Although these Mint positions might have been considered somewhat sinecures, Newton took them so seriously that he retired from Cambridge University and moved to London in 1701 so that he could properly police the English currency against counterfeiters. He found that about 20 percent

Figure 17.5. Sir Isaac Newton. (Stipple engraving of Isaac Newton, by John Vanderbank, at the National Library of Wales, based on a 1725 portrait in the collection of the Royal Society.)

of the coins produced during the great recoinage in 1696 were counterfeit. It should be noted that, at that time, counterfeiting was considered high treason and punishable by death. Newton proved quite adept at catching and prosecuting counterfeiters. This was the beginning of the end of his scientific career. During this time, he also had some psychological break-downs, under the influence of which, through written correspondence, he alienated colleagues and broke off relationships with such luminaries as John Locke. Postmortem investigations indicate that mercury poisoning might explain Newton's eccentricities in later life. In time, he recovered his senses and continued his Mint activities, which brought him a rather hand-some salary and made him a relatively rich man.

In April 1705, Newton was knighted by Queen Anne during a visit to Trinity College at Cambridge. It is still speculated today that Newton received a knighthood as a political gesture rather than as an acknowledgment of his scientific brilliance or his service as master of the Mint.

In the waning years of his life, Newton lived in Cranbury Park near Winchester, England, with his niece, Catherine Barton Conduitt, and her husband, John Conduitt. By this time, he was considered one of the most famous scientists of his day and was quite wealthy and generous to charities. A lifelong bachelor, Newton died in London on March 20, 1727; he was buried in Westminster Abbey.

CHAPTER 18

❦

Gottfried Wilhelm (von) Leibniz: German (1646–1716)

On February 15, 1946, the front page of the *New York Times* featured the announcement of "an amazing machine which applies electronic speeds for the first time to mathematical tasks hitherto too difficult and cumbersome for solution," which was one of the earliest electronic general-purpose computers. The ENIAC (Electronic Numerical Integrator and Computer) was built during wartime at the University of Pennsylvania, and it is now considered a milestone in the history of computers. It was of monstrous size, as it weighed more than 25 tons and occupied an area of more than 150 square meters (see fig. 18.1).

Among other basic electronic components, such as relays, resistors, or capacitors, it contained 20,000 vacuum tubes,[1] connected through miles of wiring and approximately 5,000,000 hand-soldered joints.[2] The "mother of all electronic computers" consumed 150,000 watts of electricity; in comparison, a modern desktop computer needs only about 200 watts. The immense power requirement led to the rumor that whenever ENIAC was switched on, lights in Philadelphia dimmed.

ENIAC was just a large collection of electronic adding machines and other arithmetic units; it lacked stored programs or an operating system that are found in modern computers. Digits were stored using ten-position ring counters, for the ten digits of our decimal system, and each digit required thirty-six vacuum tubes. Programming the machine to solve a particular problem was done by manipulating its switches and cables, and

Figure 18.1. Betty Jennings (*left*) and Fran Bilas (*right*), operating ENIAC's main control panel. (US Army photo, ARL Technical Library, 1945–1947.)

it could take several weeks to find a problem's solution. Furthermore, vacuum tubes had a limited lifetime, and if a program didn't work properly, programmers would have to crawl inside the massive structure to find bad tubes or detect bad joints. The first programmers of ENIAC were all female: Kay McNulty, Betty Jennings, Betty Snyder, Marlyn Meltzer, Fran Bilas, and Ruth Lichterman. Their work was not widely recognized, for over fifty years. In fact, historians had at first mistaken them for models posing in front of the machine.

The enormous size and power consumption of ENIAC and similar machines developed at about the same time were mainly caused by the vacuum tubes, which were used as switches and amplifiers; thus, they represented essential elements of any electronic device. Similar to today's incandescent light bulbs, vacuum tubes were bulky, produced a lot of heat, and frequently failed. The solution to this issue came in 1947, when William Shockley, John Bardeen, and Walter Brattain of Bell Laboratories discovered the transistor effect in semiconductors, allowing them to build an electric switch made of solid materials—a transistor—obviating the need for a vacuum tube. They were awarded the Nobel Prize in Physics in 1956 for their invention of the transistor, which is still considered one of the greatest breakthroughs in technology history. Transistors were much smaller and faster, and they were more reliable and more powerful than vacuum tubes. Through the late 1950s and 1960s, the so-called first-generation computers, made with bulky vacuum tubes, were replaced by second-generation computers using transistors instead. The invention of the transistor laid the foundation for the digital revolution that began in the second half of the twentieth century and continues to the present day.

Transistors are the basic building blocks of microprocessors. The central processing unit (CPU) of your computer contains billons of transistors, all packed into an area of about a square inch. A transistor is essentially an electronic switch, which can be in either an "on" or an "off" state. These two possible states or positions define what is called Boolean algebra: an algebra in which the values of the variables are the truth-values: "true" and "false," usually denoted 1 and 0. Indeed, the language of computers consists of only two symbols: zeros and ones. The basic unit of information is a binary digit or "bit" (which can be either 0 or 1). A sequence of 8 bits is called a byte; a kilobyte consists of 1,000 bytes; a megabyte of 1,000 kilobytes, and so on. Interpreting a sequence of bits as a place-value notation with consecutive powers of 2, we obtain a binary number: For example, the binary number 10001001 represents the decimal number $2^7 + 2^3 + 2^0 = 137$; likewise, the binary number 11111111 represents the decimal number $2^7 + 2^6 + 2^5 + 2^4 + 2^3 + 2^2 + 2^1 + 2^0 = 255$. Thus, with one byte of information, we can encode 256 different characters (since we can represent all numbers from 0 to 255). Modern computer architectures typically use "words" of 32 or 64 bits, built from 4 or 8 bytes. It is quite remarkable that binary numbers have already been thoroughly studied by the German philosopher and mathematician Gottfried Wilhelm Leibniz, long before 1800, when Alessandro Volta invented the first battery and built electrical circuits. Leibniz invented binary arithmetic, which is in fact used by virtually all modern computers. However, he not only laid the theoretical foundations for electronic computers but also anticipated them by describing machines that were, at least in principle, capable of solving complex mathematical problems. Moreover, he is credited, along with Isaac Newton, with the development of differential and integral calculus. And it should be said that today we use Leibniz's symbolism in calculus rather than Newton's symbolism, as was mentioned in the previous chapter. While giving a short overview of his life, we will illuminate a few of his contributions to mathematics.

Electronic Computer Flashes Answers, May Speed Engineering

By T. R. KENNEDY Jr.
Special to THE NEW YORK TIMES.

PHILADELPHIA, Feb. 14—One of the war's top secrets, an amazing machine which applies electronic speeds for the first time to mathematical tasks hitherto too difficult and cumbersome for solution, was announced here tonight by the War Department. Leaders who saw the device in action for the first time heralded it as a tool with which to begin to rebuild scientific affairs on new foundations.

Such instruments, it was said, could revolutionize modern engineering, bring on a new epoch of industrial design, and eventually eliminate much slow and costly trial-and-error development work now deemed necessary in the fashioning of intricate machines. Heretofore, sheer mathematical difficulties have often forced designers to accept inferior solutions of their problems, with higher costs and slower progress.

The "Eniac," as the new electronic speed marvel is known, virtually eliminates time in doing such jobs. Its inventors say it computes a mathematical problem 1,000 times faster than it has ever been done before.

The machine is being used on a problem in nuclear physics.

The Eniac, known more formally as "the electronic numerical integrator and computer," has not a single moving mechanical part. Nothing inside its 18,000 vacuum tubes and several miles of wiring moves except the tiniest elements of matter-electrons. There are, however, mechanical devices associated with it which translate or "interpret" the mathematical language of man to terms understood by the Eniac, and vice versa.

Ceremonies dedicating the machine will be held tomorrow night at a dinner given a group of Government and scientific men at the University of Pennsylvania, after

Column 3

Figure 18.2.

Gottfried Wilhelm Leibniz was born on the July 1, 1646, to Friedrich Leibniz, a professor of moral philosophy at Leipzig, and Catharina Schmuck, Friedrich's third wife. Friedrich Leibniz died when Gottfried was six years old. At the age of seven, Leibniz entered the Nicolai School in Leipzig and from that time on, he was also given free access to his father's personal library, which he would later inherit. His father's library seems to have been more influential on his education than the curriculum he was taught at school. Strongly motivated by wanting to read his father's books, Leibniz taught himself Latin by comparing Latin and German descriptions in illustrated books. So as to decipher the meaning of the Latin words, he compared the descriptions of the same pictures in two different books. By the age of twelve, he was proficient in the Latin language, mastering it far beyond school level. Through his father's library, he had access to advanced texts of philosophy and theology. His skill at Latin enabled him to read these books. In 1661, at the age of fourteen, he enrolled in his father's former university and completed his bachelor's degree in philosophy in December 1662. He then spent the summer term of 1663 in Jena, Germany, where Erhard Weigel was professor of mathematics. Weigel was also a philosopher who believed that numbers were the fundamental concept of the universe. Back in Leipzig, Leibniz applied the mathematical ideas he had learned from Weigel to his studies in philosophy and law. In particular, he assigned values of 0, 1, and $\frac{1}{2}$ to conditions of law that were impossible, necessary (absolute), or conditional, respectively. Leibniz was awarded his bachelor's degree in law in 1665, the year after his mother had died. He then started to work on his habilitation thesis in philosophy, which was then included as part of his first book, *Dissertatio de arte combinatoria* (*On the Combinatorial Art*). In 1666, the University of Leipzig rejected Leibniz's doctoral application, probably because of his relative youth, as compared to the other candidates, and the limited number of available tutorials. Yet Leibniz did not want to wait another year, so he went to the University of Altdorf, where he received a doctorate in law

Figure 18.3. Gottfried Wilhelm Leibniz.

in 1667 for his dissertation *De casibus perplexis* (On Perplexing Cases). He was immediately offered an academic appointment at the University of Altdorf but declined it for a position as a secretary to an alchemical society in Nuremberg, Germany. He then met Baron Johann Christian von Boyneburg, who hired him as his assistant, librarian, lawyer, and advisor. While working on various projects for Von Boyneburg, Leibniz published political essays and continued his law career, promoted by the baron, who used his personal contacts to enhance Leibniz's career. In 1672, Leibniz went to Paris on a diplomatic mission related to his role as an advisor to German authorities. In Paris, he met the Dutch physicist and mathematician Christiaan Huygens (1629–1695), who introduced Leibniz to the works of the great mathematicians of the time and mentored Leibniz in his self-study of the newest developments in mathematics. In a letter to his friend the Swiss mathematician Johann Bernoulli (1667–1748) in 1703, Leibniz wrote, "When I arrived in Paris in the year 1672, I was self-taught with regard to geometry, and indeed had little knowledge of the subject, for which I had not the patience to read through the long series of proofs."[3] Leibniz first demonstrated mathematical talent when Huygens sent him a problem for which he himself had already found the solution. The problem was to find the sum of the infinite series of reciprocal triangular numbers, which are the numbers representing objects that can be arranged in an equilateral triangle, as is shown in figure 18.4.

Let us first look at a finite series of, say, six numbers, and consider their sum, $a_1 + a_2 + a_3 + a_4 + a_5 + a_6$. Leibniz noticed that if we take any given finite series of numbers and form the series of differences of successive terms, such as:

$$d_1 = a_2 - a_1, d_2 = a_3 - a_2, d_3 = a_4 - a_3, d_4 = a_5 - a_4, d_5 = a_6 - a_5,$$

then the sum of these differences is simply the difference between the last and first terms of the original series:

$$d_1 + d_2 + d_3 + d_4 + d_5 = a_6 - a_1.$$

Thus, if we can express a given series $b_1, b_2, b_3, \ldots$ as a series of differences of successive terms of another series, $a_1, a_2, a_3, \ldots$, then we can compute the sum of the original series, $b_1 + b_2 + \ldots + b_n$, by simply subtracting the first from the last term of the other series: $b_1 + b_2 + \ldots + b_n = a_{n+1} - a_1$. To compute the sum of the reciprocals of the triangular numbers, Leibniz wrote this sum as a sum of differences, which turns out to be twice the sum of differences of the reciprocals of the natural numbers:

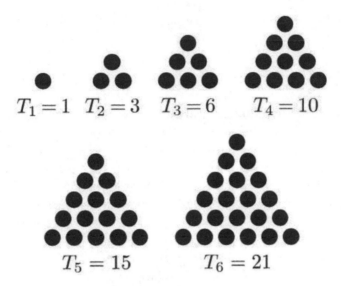

Figure 18.4.

$$\frac{1}{1}+\frac{1}{3}+\frac{1}{6}+\frac{1}{10}+\frac{1}{15}=\left(\frac{2}{1}-\frac{2}{2}\right)+\left(\frac{2}{2}-\frac{2}{3}\right)+\left(\frac{2}{3}-\frac{2}{4}\right)+\left(\frac{2}{4}-\frac{2}{5}\right)+\left(\frac{2}{5}-\frac{2}{6}\right)=\frac{2}{1}-\frac{2}{6}=\frac{5}{3}$$

Leibniz described this procedure as follows: "If one wants to add, for example, the first five fractions (reciprocals of triangular numbers) from $\frac{1}{1}$ to $\frac{1}{15}$ inclusive, one takes the number of fractions, that is, 5, added to 1 to get 6, and creates the fraction $\frac{5}{6}$, which when doubled is $\frac{10}{6}$, or $\frac{5}{3}$, which gives us the sum:

$$\frac{1}{1}+\frac{1}{3}+\frac{1}{6}+\frac{1}{10}+\frac{1}{15}$$

which is, the same as if one had added these fractions together."[4] If we denote the triangular numbers by T_1, T_2, T_3, . . ., we have thus obtained a simple formula for the sum of their reciprocals:

$$\frac{1}{T_1}+\frac{1}{T_2}+\frac{1}{T_3}+\ \ldots\ +\frac{1}{T_n}=2\cdot\left(1-\frac{1}{n+1}\right)$$

This procedure can be extended to the infinite series of triangular numbers by letting n go to infinity in the expression

$$2\cdot\left(1-\frac{1}{n+1}\right)$$

This term will approach 2 as n gets larger and larger, and, thus, the sum of the infinite series

$$\frac{1}{T_1} + \frac{1}{T_2} + \frac{1}{T_3} + \ldots$$

is equal to 2. By applying his method to other infinite series, Leibniz was able to calculate their limits as well. He also anticipated the modern mathematical definition of the sum of an infinite series as a limit, when he wrote in an unpublished paper of April 1676: "Whenever it is said that a certain infinite series of numbers has a sum, I am of the opinion that all that is being said is that any finite series with the same rule has a sum, and that the error always diminishes as the series increases, so that it becomes as small as we would like."[5]

Leibniz was soon acquainted with the mathematical achievements of his generation and began to pursue his own ideas, which would lead to significant contributions in mathematics. He spent the next four years in Paris, interrupted only by a trip to London, where he visited the Royal Society and presented a not-yet-complete calculating machine that he had designed and that would be the first one capable of executing all four basic operations (adding, subtracting, multiplying, and dividing). The members of the Royal Society were very impressed and quickly elected him as an external fellow. During his stay in Paris, he developed his variant of infinitesimal calculus using differentials, as well as most of his other substantial works in mathematics. In a manuscript dated November 21, 1675, he used his notation for the integral of a function, $\int f(x)dx$, for the first time. In a letter to Newton, he employed his method of differentials, which Newton recognized as being equivalent to his own method of fluxions. Newton had described some of his results in an earlier letter to Leibniz (yet without describing his methods) and believed that Leibniz had stolen his ideas. In fact, Newton employed his equivalent formulation of calculus using fluxions as early as 1666 but did not publish it until 1693. Newton rightfully claimed that not a single previously unsolved problem was solved by Leibniz's approach to calculus. While this is true, as we mentioned earlier, Leibniz must still be credited for developing a modern mathematical notation that is now standard in calculus, whereas Newton's impractical notation was eventually abandoned. During his time in Paris, Leibniz absorbed all contemporary mathematics and reformulated it in an improved system of notation, thereby simplifying calculations and making the tools of calculus much more easily accessible for other mathematicians and physicists. Although he did not obtain any new mathematical results with his variant of calculus, his superior system of notation proved to be highly influential for further developments in mathematics. Leibniz would have liked to have remained in Paris and tried to become an honorary member of the French Academy of Sciences, but no invitation came. His patron, von Boyneburg, had died; without a professional perspective in Paris, in October 1676, he therefore accepted a position as a librarian and court councilor at the House of Hanover. The House of Hanover is a German royal dynasty, formally named the House of Brunswick-Lüneburg, Hanover line. It provided monarchs to Great Britain and Ireland and ruled the United Kingdom of Great Britain throughout the nineteenth century. Around 1679, Leibniz developed binary arithmetic, which was published in his 1703 article "Explication de l'arithmétique binaire" (Explanation of Binary Arithmetic). In the introduction, he writes:

> The ordinary reckoning of arithmetic is done according to the progression of tens. Ten characters are used, which are 0, 1, 2, 3, 4, 5, 6, 7, 8, 9, which signify zero, one, and the successive numbers up to nine, inclusively. And then, when reaching ten, one starts again, writing ten as "10," ten times ten, or one hundred, as "100," ten times one hundred, or one thousand, as "1000," ten times one thousand by "10000," and so on. But instead of the progression of tens, I have for many years used the simplest progression of all, which proceeds by twos, having found that it is useful for the perfection of the science of numbers. Thus, I use no other characters in it except 0 and 1, and when reaching two, I start again. This is why two is here expressed by "10," and two times two, or four, by "100," two times four, or eight, by "1000," two times eight, or sixteen, by "10000," and so on.[6]

As we have done earlier, to convert a binary number into a decimal number, we just have to add the represented powers of 2. For example, the binary number 101011 represents (read from right to left) $1 \cdot 2^0 + 1 \cdot 2^1 + 0 \cdot 2^2 + 1 \cdot 2^3 + 0 \cdot 2^4 + 1 \cdot 2^5 = 43$. To convert a decimal number into its binary representation, we divide by 2, and write down the remainder R (which must be 0 or 1), and repeat this procedure until we reach 0:

$43 \div 2 = 21$ R: 1
$21 \div 2 = 10$ R: 1

$10 \div 2 = 5$ R: 0
$5 \div 2 = 2$ R: 1
$2 \div 2 = 1$ R: 0
$1 \div 2 = 0$ R: 1

The sequence of remainders, reading from bottom to top, gives us the binary representation of 43 as 101011. Why does this work?

Dividing the expression $1 \cdot 2^0 + 1 \cdot 2^1 + 0 \cdot 2^2 + 1 \cdot 2^3 + 0 \cdot 2^4 + 1 \cdot 2^5 = 43$ by 2 reduces by 1 the exponents of all powers of 2, yielding $1 \cdot 2^0 + 0 \cdot 2^1 + 1 \cdot 2^2 + 0 \cdot 2^3 + 1 \cdot 2^4$ with a remainder of 1. Again, dividing by 2, we arrive at $0 \cdot 2^0 + 1 \cdot 2^1 + 0 \cdot 2^2 + 1 \cdot 2^3$ with a remainder of 1; and, continuing this process, we obtain all digits of the binary representation of 43, the last one being the one with the highest place value, and, therefore, the final remainder after successive divisions by 2.

After providing conversion tables for the numbers from 0 to 32, Leibniz explains addition, subtraction, multiplication, and division for binary numbers. These operations are actually much like the normal decimal operations with which we are familiar, except that they carry a value of 2 instead of a value of 10. For example, adding the decimal numbers 5 and 8 gives a last digit of 3, and we carry 1 to the next digit to obtain 13. In the same fashion, adding the binary numbers 1 and 1, we get 0 as the last digit and carry 1 to the next digit, giving the binary number 10_2 (here we place the subscript 2 to distinguish binary numbers from decimal numbers). To add the binary numbers 110_2 and 111_2, we start from the last digit: Adding 0 and 1, we get 1 (with no carry). Thus, the last digit of the answer will be 1. We then move one digit to the left: Adding 1 and 1 gives us 0 and a carry of 1. The next digit of the answer will be 0, and we have to a carry a 1. Moving on to the next digit, we add 1 and 1 and the carry from the last digit, which gives us 1 and a carry of 1. Hence, we get the answer 1101_2. Figure 18.5 shows additional examples from Leibniz's original work "Explication de l'arithmétique binaire" (he wrote in French because the article was published by the French Academy of Sciences). You may want to try binary subtraction, multiplication, and division as well.

Is his article, Leibniz then argues that binary arithmetic is actually simpler than decimal arithmetic, since one doesn't have to memorize any addition or multiplication tables; it suffices to know how to add and multiply zeros and ones. As he goes on, he writes:

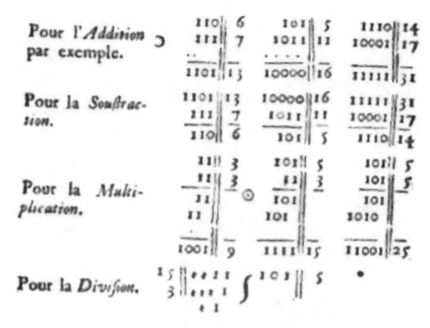

Figure 18.5. "Explication de l'arithmétique binaire."
(Academie royale des sciences, 1703.)

However, I am not in any way recommending this way of counting as an effort to introduce it in place of the ordinary practice of counting by ten. For, aside from the fact that we are accustomed to this, we have no need to learn what we have already learned by heart. The practice of counting by ten is shorter and the numbers not as long. And if we were accustomed to proceed by twelves or sixteens, there would be even more of an advantage. But calculating by twos, that is, by 0 and 1, as compensation for its length, is the most fundamental way of calculating for science, and provides for new discoveries, which are then found to be useful, even for the practice of numbers and especially for geometry. The reason for this is that, as numbers are reduced to the simplest form, such as 0 and 1, a wonderful order is apparent throughout.[7]

In fact, Leibniz already imagined a machine in which binary numbers were represented by marbles, governed by a rudimentary sort of punched

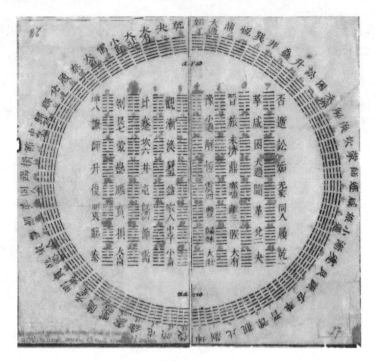

Figure 18.6. Diagram of hexagrams sent by Joachim Bouvet (1656–1730), a French Jesuit who traveled extensively in China, to Leibniz, 1701. (Franklin Perkins, *Leibniz and China: A Commerce of Light* (Cambridge: Cambridge University Press, 2004.)

cards, thereby anticipating the principle of modern electronic digital computers.

Leibniz spent the rest of his life in Hanover, except for some extensive travels through Europe. Besides his duties at the court, he kept writing about mathematics, logic, physics, and philosophy, and he held correspondence with many of the great scholars of his time. He wrote in several languages, primarily in Latin, French, and German. Moreover, Leibniz was perhaps the first major European intellectual with a close interest in Chinese civilization, and he noted with fascination that the hexagrams in the I Ching, an ancient Chinese divination text, correspond to the binary numbers from 000000 to 111111. Figure 18.6 shows a diagram of the I Ching hexagrams, with Arabic numerals added by Leibniz.

Apart from his contributions to mathematics, he is also remembered as a major figure in philosophy and logic, and as the inventor of the step

Figure 18.7.

calculator and a digital mechanical calculator that he developed in 1694 (see fig. 18.7). Together with René Descartes and Baruch Spinoza, Leibniz was one of the three great rationalists of the seventeenth century, forerunners of the Age of Enlightenment. He was one of the last polymaths and left an incredible number of writings and letters on diverse subjects, some 200,000 pages of written and printed paper have survived. That his broad interests and richness of ideas might sometimes have been a burden is revealed in a letter he wrote to a friend, indicating his slight desperation about the limited amount of time available for his intellectual pursuits:

> I cannot tell you how extraordinarily distracted and spread out I am. I am trying to find various things in the archives; I look at old papers and hunt up unpublished documents. From these I hope to shed some light on the history of the [House of] Brunswick. I receive and answer a huge number of letters. At the same time, I have so many mathematical results, philosophical thoughts, and other literary innovations

that should not be allowed to vanish, and that I often do not know where to begin.[8]

The last years of Leibniz's life were embittered by a long controversy with Newton, and others, over whether Leibniz had discovered calculus independently of Newton, or whether he had merely invented another notation for ideas that were fundamentally Newton's. He received little appreciation for his mathematical works during his lifetime and died lonely (as he never married) in Hanover in 1716. Although Leibniz was a life member of the Royal Society and the Berlin Academy of Sciences, neither organization saw fit to honor his death. In 1985, the German government created the Leibniz Prize, one of the world's largest prizes for scientific achievement.

Chapter 19

Giovanni Ceva:
Italian (1647–1734)

It is not uncommon that a person of outstanding genius is remembered largely as a result of one work. This trend is particularly noticeable in the fields of music and literature. It is also the case in mathematics. Here we have a mathematician whose primary area of interest was geometry, but who also dabbled to some extent in economics and physics. This was the Italian mathematician Giovanni Ceva who was born in Milan, Italy, on September 1, 1647, into what would be considered in his day a wealthy family. Although we know very little about his early years, Ceva later made comments about his youth being sad, with many kinds of misfortune. He received his education at the Jesuit College in Milan, at the Palazzo di Brera, where he showed an early talent for mathematics and science.

After completing his college studies, he followed his father's footsteps, engaging in business and political activities largely in Milan, Genoa, and Mantua, yet his interest in science and mathematics continued. He eventually entered the University of Pisa in 1670 to study mathematics. During his time at the university, he struggled with the problem of squaring the circle, that is, of constructing a square with the same area as a given circle using only a straight edge and a pair of compasses. Rather discouraged, he gave up this effort; interestingly, in 1882, the problem of squaring the circle was proved to be impossible, as a result of the Lindemann-Weierstrass theorem, which proved that π is a transcendental number. In 1678, Ceva

Figure 19.1. Giovanni Ceva.

published his book *De lineis rectis se invicem secantibus statica constructio*, while continuing to work with his father. This work took him to Mantua and Montferrat in a position such that he was responsible for the economy of these two states. This responsibility did not deter him from further pursuing mathematical interests and publishing other works. His most effective employment was rewarded by the Duke of Mantua, who granted Ceva citizenship there. While pursuing his mathematical studies alongside his work, he corresponded with many of the leading mathematicians of his day.

He married Cecilia Vecchi on January 15, 1685. They had seven children, and the five that survived birth were also given citizenship by the Duke of Mantua. In 1686, he was appointed professor of mathematics at the University of Mantua. He continued at this university position for the rest of his life, including the time, in 1707, when the region was under Austrian protection.

Ceva is remembered today for a theorem about triangles that he published in his 1678 book, *De lineis rectis se invicem secantibus statica constructio*. Ceva's theorem states that the product of the alternate segments along the sides of the triangle, determined by the intersections of concurrent line segments (called *cevians*) emanating from the triangle's vertices and ending at the opposite side, are equal.

Ceva's theorem is an equivalence, or biconditional, which means that the converse is also true. To justify it requires two proofs—the original statement and its converse. Ceva's theorem states the following, referring to triangle *ABC* shown in figure 19.2: The three lines containing the vertices of triangle *ABC* and intersecting the opposite sides in points *L*, *M*, and *N*, respectively, are concurrent if and only if $AM \cdot BN \cdot CL = MC \cdot NA \cdot BL = 1$, or, stated another way, $\frac{AM}{MC} \cdot \frac{BN}{NA} \cdot \frac{CL}{BL} = 1$.

There are many proofs available to justify this theorem, yet we shall use just one of these methods to prove this wonderful theorem.[1] It is perhaps easier to follow the proof by looking at the left-side diagram in figure 19.2, and then verifying the validity of each of the statements in the right-side diagram. In any case, the statements made in the proof hold for both diagrams.

In figure 19.2 we have on the left, triangle *ABC*, with a line (*SR*) containing *A* and parallel to *BC*; *SR* intersects *CP* extended at *S* and *BP* extended at *R*.

The parallel lines enable us to establish the following pairs of similar triangles:

$$\triangle AMR \sim \triangle CMB; \text{ therefore, } \frac{AM}{MC} = \frac{AR}{CB}. \tag{I}$$

$$\triangle BNC \sim \triangle ANS; \text{ therefore, } \frac{BN}{NA} = \frac{CB}{SA}. \tag{II}$$

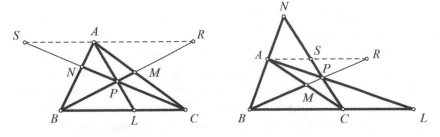

Figure 19.2.

$\triangle CLP \sim \triangle SAP$; therefore, $\dfrac{CL}{SA} = \dfrac{LP}{AP}$.   (III)

$\triangle BLP \sim \triangle RAP$; therefore, $\dfrac{BL}{RA} = \dfrac{LP}{AP}$.   (IV)

From (III) and (IV) we get: $\dfrac{CL}{SA} = \dfrac{BL}{RA}$

This can be rewritten as $\dfrac{CL}{BL} = \dfrac{SA}{RA}$ (V)

Now by multiplying (I), (II), and (V) we obtain our desired result:

$$\frac{AM}{MC} \cdot \frac{BN}{NA} \cdot \frac{CL}{BL} = \frac{AR}{CB} \cdot \frac{CB}{SA} \cdot \frac{SA}{RA} = 1$$

This can also be written as $AM \cdot BN \cdot CL = MC \cdot NA \cdot BL$.

The converse of this theorem is of particular value as well. That is, if the products of the alternate segments along the sides of the triangle are equal, then the cevians determining these points must be concurrent.

We shall now prove that if the lines containing the vertices of triangle ABC intersect the opposite sides in points L, M, and N, respectively, so that $\frac{AM}{MC} \cdot \frac{BN}{NA} \cdot \frac{CL}{BL} = 1$, then these lines AL, BM, and CN, are concurrent.

Suppose BM and AL intersect at P. Draw PC and call its intersection with AB point N'. Now that AL, BM, and CN' are concurrent, we can use the part of Ceva's theorem proved earlier to state the following:

$$\frac{AM}{MC} \cdot \frac{BN'}{N'A} \cdot \frac{CL}{BL} = 1$$

But our hypothesis stated that

$$\frac{AM}{MC} \cdot \frac{BN}{NA} \cdot \frac{CL}{BL} = 1$$

Therefore, $\dfrac{BN'}{N'A} = \dfrac{BN}{NA}$, so that N and N' must coincide, which thereby proves the concurrency.

For convenience, we can restate this relationship as follows: If $AM \cdot BN \cdot CL = MC \cdot NA \cdot BL$, then the three lines are concurrent.

There is an interesting variation to Ceva's theorem, which was discovered by the French mathematician Lazare Carnot (1753–1823).[2] Here the concurrency of the three cevians is specified by the partitioned angles at the triangle's vertices. In figure 19.3, we have cevians AL, MB, and CN that partition the angles as follows:

angle A is partitioned into α_1 and α_2,

angle B is partitioned into β_1 and β_2, and

angle C is partitioned into γ_1 and γ_2.

They will meet at a common point P if and only if[3] $\frac{\sin\alpha_1}{\sin\alpha_2} \cdot \frac{\sin\beta_1}{\sin\beta_2} \cdot \frac{\sin\gamma_1}{\sin\gamma_2}$ = 1. The proof entails using the law of sines several times and is left to the ambitious reader.

Ceva further enriched our knowledge of geometry by discovering a theorem that was developed in 100 CE by the Greek mathematician Menelaus of Alexandria (70–140 CE),[4] who established that the equal products of alternate segments on the sides of a triangle determine collinear points as you can see from the following statement of Menelaus's theorem:

> If three points, X, Y, and Z, are located on the sides (or their extensions, as shown in fig. 19.4b) of triangle ABC such that $AZ \cdot BX \cdot CY = AY \cdot BZ \cdot CX$, then the three points X, Y, and Z are collinear. (See figs. 19.4a and 19.4.b.)

The proof of Menelaus's theorem is rather straightforward and once again uses elementary geometry relationships. Once again, this is a biconditional relationship and can be stated symbolically as follows:

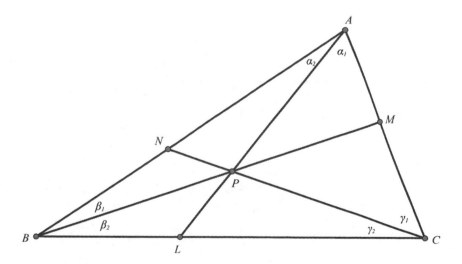

Figure 19.3.

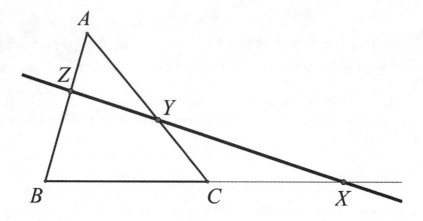

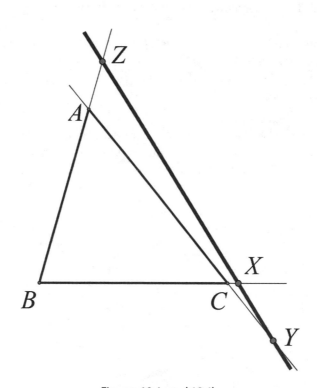

Figures 19.4a and 19.4b.

$AZ \cdot BX \cdot CY = AY \cdot BZ \cdot CX$ if and only if X, Y, and Z are collinear.

We shall first prove that if X, Y, and Z are collinear, then $AZ \cdot BX \cdot CY = AY \cdot BZ \cdot CX$.

Draw a line containing C, parallel to AB, and intersecting XYZ or YXZ at D, as shown in figure 19.5. We are thus beginning with the given collinear points X, Y, and Z.

For $\triangle CDX \sim \triangle BXZ$, therefore, $\dfrac{CD}{BZ} = \dfrac{CX}{BX}$, or $CD = \dfrac{BZ \cdot CX}{BX}$

$$\frac{CD}{BZ} = \frac{CX}{BX}, \text{ or } CD = \frac{BZ \cdot CX}{BX} \tag{I}$$

For $\triangle CDY \sim \triangle APZ$, therefore, $\dfrac{CD}{AZ} = \dfrac{CP}{AY}$, or $CD = \dfrac{AZ \cdot CP}{AY}$

$$\frac{CD}{AZ} = \frac{CP}{AY}, \text{ or } CD = \frac{AZ \cdot CP}{AY} \tag{II}$$

From equations (I) and (II): $\dfrac{BZ \cdot CX}{BX} = \dfrac{AZ \cdot CP}{AY}$

$$\frac{BZ \cdot CX}{BX} = \frac{AZ \cdot CP}{AY},$$

from which we easily get $AZ \cdot BX \cdot CY = AY \cdot BZ \cdot CX$.

Now we shall prove that if the points X, Y, and Z are so situated that the equation $AZ \cdot BX \cdot CY = AY \cdot BZ \cdot CX$ is true (or another way of expressing this is that $\frac{AY}{CY} \cdot \frac{BZ'}{AZ'} \cdot \frac{CX}{BX} = 1$), then the three points X, Y, and Z are collinear.

We will let the intersection point of AB and XY be the point Z'. Then we have to prove $Z' = Z$.

Because of part 1 (above) we have $\frac{AY}{CY} \cdot \frac{BZ}{AZ} \cdot \frac{CX}{BX} = 1$, also $\frac{BZ'}{AZ'} = \frac{BZ}{AZ}$.

Therefore, we have $Z' = Z$, and the points X, Y, and Z must be collinear.

The sad thing about Ceva's fame today is that during his time, his 1678 book was not popular; it was only published in one edition. Many of his findings were then later discovered by other mathematicians, who were not aware of Ceva's work, and so they were unable to attribute their findings to him. It was not until the nineteenth century that the French mathematician Michel Chasles (1793–1880) recognized that Ceva's work predated those of subsequent mathematicians. Therefore, we refer to this theorem now correctly as Ceva's theorem. Beyond his work in geometry, Ceva also worked with applications of mechanics, but clearly his fame today is limited to his work in geometry. Ceva died in Mantua, Italy, on May 13, 1734.

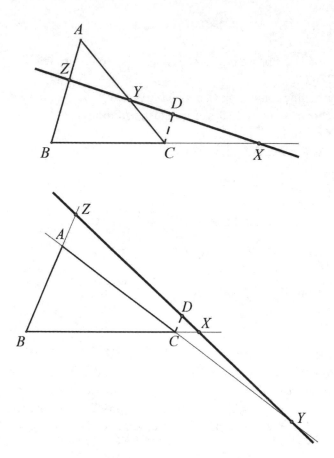

Figure 19.5.

CHAPTER 20

Robert Simson:
Scottish (1687–1768)

The United States education system is one of the few in the world where students study one year of geometry while still in high school. Naturally, the curriculum for this course is ultimately based on the famous work by Euclid, *Elements*. Yet the path the high-school geometry curriculum took is rather interesting. We could begin with the Scottish mathematician Robert Simson, who set out to prepare a perfect text, in English, of Euclid's first six books together with the eleventh and twelfth books, and first published it as a complete book in Glasgow, Scotland, in 1756. We show the title page of the 1787 edition in figure 20.1.

In 1794, the French mathematician Adrien-Marie Legendre (1752–1833) published a geometry book titled Éléments *de géométrie*, which remained popular for about one hundred years. This book made its way across the Atlantic Ocean through the work of the American mathematician Charles Davies, whose adaptation, *Elements of Geometry* was first published in 1828. This title then became the model for teaching geometry in the United States (see fig. 20.2). Naturally, there have been many modifications over the years until we get to today's American high-school geometry course.[1]

That this path began with Robert Simson is only one aspect of this important mathematician's biography. The eldest son of John Simson, he was born on October 14, 1687, in West Kilbride, Ayrshire, Scotland.

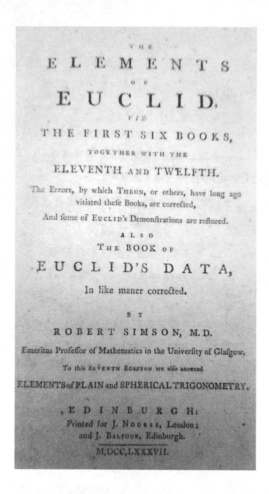

Figure 20.1.

His brilliance in classics and botany enabled him to enter the University of Glasgow at the age of fourteen, in 1702. At the urging of his father, Simson was being guided to prepare for a life in the church; however, he found the religious thinking unsatisfying, because of its speculation and lack of precision. A book on Asian philology was his next attraction, but there, even though statements could be shown to be either true or false, he was not totally satisfied. It was not until he delved into mathematics that he found a subject of particular interest. In particular, it was geometry that was his great awakening, specifically when he read Euclid's *Elements*. Unfortunately, the challenges in the field of mathematics in Glasgow were not sufficient, so he decided to go to London upon completion of his studies at the University of Glasgow in 1710, despite having been offered the chair of mathematics at the University of Glasgow. In London, he met the leading mathematicians

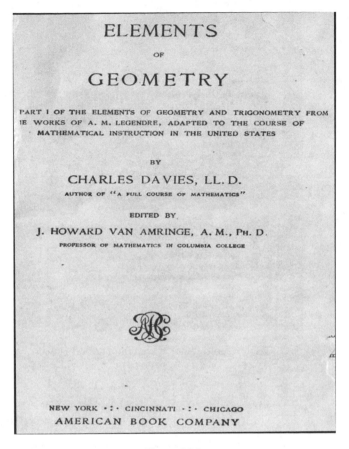

Figure 20.2.

of his time, and a year later he returned to Glasgow to assume the position that he had been originally offered. There, he once again pursued his favorite aspect of mathematics: geometry. After writing about Euclid's work in Latin, he eventually published what was probably the first English version of this monumental work by Euclid. As we mentioned earlier, Simson's book, which had more than seventy editions throughout the world, still remains as the primary model for studying Euclid's *Elements* through today's American high-school geometry course. Throughout his career, Simson was aware of developments in algebra and infinitesimal calculus, but he usually preferred to present his mathematical ideas in terms of geometry.

Robert Simson never married and lived a very simple lifestyle, giving up a more elegant home for a small, modest apartment, and he ate most of his meals at a small pub near the university. In 1746, he was given an

Figure 20.3. Eighteenth-century portrait of Robert Simson.

honorary doctorate of medicine by St Andrews University in Scotland. He carried this degree with pride, as you can see in the title page shown in figure 20.1.

In 1753, Simson provided another aspect to our growing history of mathematics when he indicated that the ratio between adjacent Fibonacci numbers approaches the golden ratio as a limit, namely,

$$\frac{1+\sqrt{5}}{2} = 1.618\ldots \ .$$

However, his fame today beyond the historic achievement with his geometry book is a geometric theorem that bears his name, yet one that should not be attributed to him, since it was discovered by the Scottish mathematician William Wallace (1768–1843), who published it in 1799, in Thomas Leybourn's *The Mathematical Repository*. It is believed that because of Simson's popularity, geometric ideas that were not attributed to René Descartes and were written in the Euclidean style were automatically attributed to Robert Simson.

Let's take a look at this famous Simson's theorem. In figure 20.4, we notice that triangle *ABC* is inscribed in a circle, and point *P* is any point on the

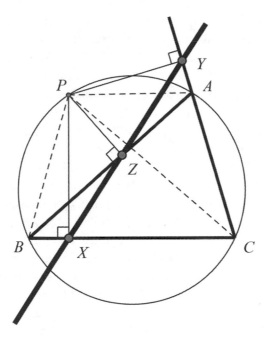

Figure 20.4.

circle. From point *P*, three perpendiculars are drawn to the three sides of the triangle (thin, solid lines). The feet of the perpendiculars are noted as the points *X*, *Y*, and *Z*. According to Simson's theorem (or should we have said "Wallace's theorem"?), these three points, *X*, *Y*, and *Z*, will always be collinear.

Justifying why this amazing relationship actually holds true is a fine exercise in understanding the power of geometric relationships. We begin by referring to figure 20.4, where we notice that angle *PYA* is supplementary to angle *PZA* (since both are right angles). We recall that when the opposite angles of a quadrilateral are supplementary, the quadrilateral is cyclic (i.e., inscribable in a circle). Therefore, quadrilateral *PZAY* is cyclic. We now draw *PA*, *PB*, and *PC* (dashed lines). Considering the circumscribed circle about quadrilateral *PZAY* (not shown), we have two angles ∠*PYZ* = ∠*PAZ* = ∠*PAB*, intercepting the same arc, namely, the arc *PZ* of the circle, which makes these two angles (∠*PAZ* = ∠*PAB*) equal.

In a similar way, we notice the two right angles, ∠*PYC* and ∠*PXC*, are supplementary, and this establishes that we have a cyclic quadrilateral *PXCY*. Therefore, as before ∠*PYX* = ∠*PCB*, since they are both measured by the arc *PX*.

Now from the cyclic quadrilateral *PACB* we have ∠*PAZ* (or ∠*PAB*) = ∠*PCB* . From the three angle equalities that we have just established, we can tie them together and obtain ∠*PYX* = ∠*PCB* = ∠*PAZ* = ∠*PYZ*, or, simply written, we have ∠*PYX* = ∠*PYZ*, which then implies the points *X*, *Y*, and *Z* are collinear. Thus, we have proved Simson's theorem. It should be noted that the converse of this is also true.

Besides collinearity, the lengths of the perpendiculars also form a relationship that merits noting. In figure 20.5, point *P* is on the circumcircle of triangle *ABC*, where perpendiculars *PX*, *PY*, and *PZ* are drawn to sides *AC*, *AB*, and *BC*, respectively. The interesting relationship that evolves is that *PA·PZ =PB·PX*. In order to justify this surprising relationship, we will identify two cyclic quadrilaterals, namely, quadrilateral *PYZB* and quadrilateral *PXAY*. The quadrilateral *PYZB* is cyclic because ∠*PYB* and ∠*PZB*, which are right angles, are both subtended by the side *PB*, and we should know that when a side of a quadrilateral subtends equal angles at two opposite vertices, then the quadrilateral is cyclic. Therefore, ∠*PBY* = ∠*PZY*. We can make a similar argument for the quadrilateral *PXAY* to be cyclic, since ∠*PXA* = ∠*PYA* = 90°. Once again, we can conclude that ∠*PXY* = ∠*PAY*. Since we

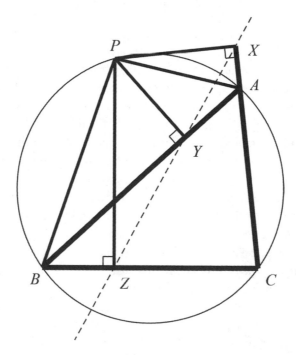

Figure 20.5.

have the points X, Y, and Z along the Simson line, we can establish that $\triangle PAB \sim \triangle PXZ$. It then follows that $\dfrac{PA}{PX} = \dfrac{PB}{PZ}$, which then gives us $PA \cdot PZ = PB \cdot PX$. This is what we set out to show in the first place.

In figure 20.6, we will show another interesting feature about the Simson line—applying it to triangle ABC. This curious relationship shows that if the altitude AD of triangle ABC meets the circumscribed circle at point P, then the Simson line (XDZ) of point P with respect to triangle ABC is parallel to the tangent, AG, to the circle at point A.

In order to show that this relationship is true, we begin by considering that the line segments PX and PZ are perpendicular, respectively, to sides AC and AB of triangle ABC. As shown in figure 20.6, we now draw the segment PB. Focusing on quadrilateral $PDBZ$, we notice that $\angle PDB = \angle PZB = 90°$, thereby allowing us to establish it as a cyclic quadrilateral. This then enables us to establish that $\angle DZB = \angle DPB$, since they both have a measure one-half of arc DB.

When we consider the circumcircle of triangle ABC, we notice two angles of equal measure, because each has a measure one half of arc AB; that is,

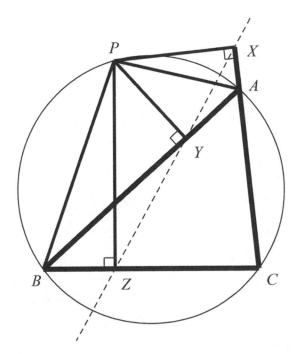

Figure 20.6.

$$\angle GAB = \frac{1}{2}\big(\text{arc}AB\big) = \angle APB \text{ (or } DPB)$$

or simply put, $\angle GAB = \angle DPB$. This, then, allows us to establish that $\angle DZB$ = $\angle GAB$, which are alternate-interior angles of the two parallel lines, AG and XDZ, formed by the transversal ABZ. Therefore, the Simson line is parallel to the tangent at point A. Even though it is believed today that Simson was not responsible for developing this theorem and its various relationships, since both it and the line bear his name, it is presented here to give a complete picture of why Robert Simson is still known very well today.

Robert Simson died in 1768, and he is buried in Blackfriars burial ground, where, sometime later, a fifty-foot monument was erected in his honor at the West Kilbride cemetery. It bears the following inscription: "The Restorer of Grecian Geometry, and by his Works the Great Promoter of its Study in the Schools."[2] This certainly summarizes his contributions for the future!

~

CHAPTER 21

Christian Goldbach: German (1690–1764)

In almost every field of endeavor, there are brilliant people who have remained famous to this day, largely because of one sterling success in their career. For example, the French composer Georges Bizet (1838–1875) is famous today for his opera *Carmen*. The American author J. D. Salinger (1919–2010) is largely remembered for his novel *The Catcher in the Rye*. Then there is the composer Engelbert Humperdinck (1854–1921), whose opera *Hänsel und Gretel* keeps his name current today. And so it is with Christian Goldbach, who is largely quoted today for the conjecture that bears his name and continued to challenge mathematicians for centuries.

Christian Goldbach (see fig. 21.1) was born on March 18, 1690, in the city of Königsberg, which was part of Brandenburg-Prussia (today it is Kaliningrad, Russia).[1] His father was a pastor in the Protestant church there, and Christian studied at the Royal Albertus University in the same city. He studied law and medicine as well as delving a bit into some mathematics. From 1710 until 1724, he traveled throughout Europe, visiting the German states, Holland, Italy, England, and France. During his visits, he sought to meet the leading scientists. For example, in 1711, he met the German mathematician and philosopher Gottfried Wilhelm Leibniz with whom he carried on a correspondence—written in Latin—for the next two years.

In 1712, he met a few mathematicians in London, including Nicolaus (I) Bernoulli and Abraham de Moivre, and he was also referred to Jacob Bernoulli. These encounters began to motivate him toward the field of mathematics. His increased interest in mathematics was brought about by meeting Nicolaus (II) Bernoulli in Venice in 1721, who then connected him with his younger brother Daniel Bernoulli, with whom a correspondence continued for another seven years. It should be noted that Goldbach was multilingual; consequently, his diary was written in German and in Latin, and his letters were written in German, Latin, French, and Italian. He also had command of the Russian language for legal documents.

In 1724, when Goldbach returned to Königsberg, he met with Georg Bernhard Bilfinger and Jakob Hermann. These two mathematicians had a great influence on Goldbach's pursuit of mathematics. His reputation in the field rose rapidly, and in 1725, Goldbach was offered a position as professor of mathematics and history at the St. Petersburg Academy of Sciences. This resulted from a reputation he built over years—reading articles by the famous mathematicians and then producing his own, which, in retrospect, were not terribly profound.

Goldbach held the position of recording secretary from the opening of the Academy of Sciences in December 1725 until January 1728. During this time, he was able to navigate the political circumstances in St. Petersburg, which was Russia's capital at that time.

In 1727, the famous and prolific Swiss mathematician Leonhard Euler arrived in St. Petersburg, where he met Goldbach, who shortly thereafter, in 1729, moved to Moscow but continued a correspondence with Euler that lasted another thirty-five years. Goldbach's role as a tutor to the royal family brought him back to St. Petersburg in 1732, where he once again became active in the Academy of Sciences. He actually was one of two people—the other being J. D. Schuhmacher—who were responsible for the administration of the Academy, and, consequently, he became increasingly more involved in the Russian government. His language competence—Latin, French, German, as well as Russian—allowed him to continuously increase his importance in the Russian government. In 1740, he resigned from the academy and was appointed to an important position in the Ministry of Foreign Affairs. In 1760, he was asked to establish a program of education for the royal family, which remained in effect for the next hundred years.

So, from where did Goldbach's legacy in mathematics evolve? As we mentioned earlier, Goldbach carried on a regular correspondence with Euler. In 1742, he proposed a conjecture to Euler that every integer greater than 2 can be expressed as a sum of three prime numbers. For example, $3 = 1 + 1$

+ 1; 4 = 1 + 1 + 2; 5 = 1 + 1 + 3; 31 = 23 + 7 + 1, and so on. It should be noted that Goldbach considered the number 1 to be a prime number, while today it is no longer considered a prime number. Euler responded to Goldbach with an equivalent form of this conjecture that all even integers greater than 2 can be expressed as the sum of two prime numbers. For example, 6 = 3 + 3; 8 = 3 + 5; 10 = 5 + 5; 31 = 29 + 2, and so on. Euler further stated that he was quite certain that this conjecture is true, but he was not able to prove it. Nor has anyone else for the past few centuries!

It is this conjecture that makes Goldbach famous still today, since no one has ever proved that it is a correct conjecture; but, on the other hand, no one has ever proved that it is not a correct conjecture, either. As a matter of fact, in 2012, the Portuguese professor Tomás Oliviera e Silva has shown the conjecture to be true for all integers less than 4,000,000,000,000,000,000, or, written succinctly, $4 \cdot 10^{18}$.[2] Today we write Goldbach's conjecture as follows: Every integer greater than 5 can be written as a sum of three primes—not including the number 1 as a prime.

Figure 21.1. Christian Goldbach.

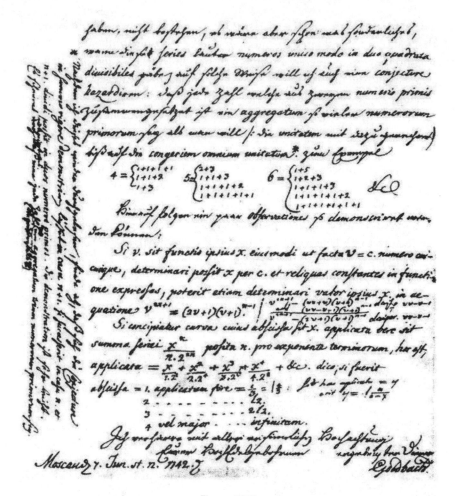

Figure 21.2.

It would appear that Goldbach considered mathematics somewhat recreational, and he motivated his correspondence partner, Euler, to test these conjectures to the first 2,500 numbers, and finding no fault.

It should be noted that a weak version of this conjecture, namely, that all odd numbers greater than 7 are the sum of three odd primes, seems to have been proven by Harald Andres Helfgott, in 2013.[3]

Goldbach's life ended on November 20, 1764, in Moscow, when he was seventy-four years old. His legacy, as we said earlier, resides with his famous conjecture, which to this day has never been proved true for all numbers, thus, it remains a conjecture and not a theorem.

CHAPTER 22

The Bernoullis:
Swiss (1700–1782)

It is perhaps unique in the history of mathematics that so many members of one family have enjoyed distinguished careers in mathematics. This is precisely the case with the Bernoulli family. Here we will focus on Jacob Bernoulli, his younger brother Johann Bernoulli, and Johann's son Daniel Bernoulli.

We begin with the eldest of the three family mathematicians whom we will highlight here: Jacob Bernoulli, who was born on December 27, 1654, in Basel, Switzerland, to Nicolaus (aka Niklaus, 1623–1708) and Margaretha Schönauer.[1] As was not unusual for the time, parents wanted their children either to prepare to take over a family business or to enter the ministry. In Jacob's case, it was the latter. Following the guidance of his parents, he studied philosophy and theology at the University of Basel, earning a master's degree in philosophy in 1671 and a second degree in theology, in 1676. Despite these official studies, Jacob secretly, and against his parents' wishes, studied mathematics and astronomy privately. Although he may have eased the family's dislike for the study of mathematics, he was not alone in this sort of familial pressure; the other mathematicians seemed to have faced similar pressures. Between the ages twenty-two and twenty-eight, Jacob traveled throughout Europe, visiting the leading scientists and mathematicians of his time.

Returning to the University of Basel in 1683, he began teaching mechanics as it relates to liquids and solids. Having a degree in theology

Figure 22.1. Jacob Bernoulli. (Painting by Jacob Bernoulli's
brother Nicolaus Bernoulli [1662–1716], 1687.)

qualified him for an appointment in the church, which he promptly turned
down so that he could pursue his true areas of interest: mathematics and
science. In 1684, he married Judith Stupanus, with whom he had two children. He continued to correspond with mathematicians and was particularly fascinated with the works of René Descartes and Gottfried Wilhelm
Leibniz. Jacob Bernoulli provided the developing subject of mathematics
with one of its more significant cornerstones: the natural logarithm,

$$e = \lim_{n \to \infty} \left(1 + \frac{1}{n}\right)^n,$$

which evolved from his study of compound interest. He found that if a
man invests $1.00 and pays 100 percent interest per year, at the end of the
year, the value is $2.00. If the interest is computed and added twice in the
year, then the $1 is multiplied by 1.5 twice, yielding $1.00 · 1.5² = $2.25. By

compounding quarterly, the investment yields $1.00 \cdot 1.25^4 = \$2.4414 \ldots$, and by compounding monthly, the investment yields $\$1.00 \cdot (1.0833 \ldots)^{12}$ $= \$2.613035 \ldots$. By compounding weekly, he found it to be $\$2.692597 \ldots$.; compounding daily yields $\$2.714567 \ldots$. With ever more frequent compounding, the amount will reach $\$2.718281828459 \ldots$ as a limit. This is the natural logarithm, and it is designated by the letter e, a symbol that was popularized by Leonhard Euler.

In 1687, he finally settled down and accepted the position of professor of mathematics at the University of Basel. This also gave him further opportunity to continue to investigate discoveries by the English mathematicians Isaac Barrow (1630–1677) and John Wallis (1616–1703), and which motivated him to further study infinitesimal geometry.

Having now secured the position at the University of Basel, he had also begun to tutor his younger brother Johann in mathematics. The brothers became interested in Leibniz's 1684 paper on differential calculus, which was published in his *Acta Eruditorum*, and might be considered as the first appearance of calculus as we know it. However, at about the same time, Isaac Newton also developed the concept of calculus, which he called fluxions. It is well known that a controversy evolved regarding who was the first to come up with this seemingly new field of mathematics. Of course, the Bernoulli brothers supported Leibniz. Today we use Leibniz's subject title of "differential calculus" as well as his symbols. In a paper he published in 1690, Jacob Bernoulli was the first to use the term "integral calculus" when analyzing a curve. In 1691, Jacob wrote about the catenary curve, which is a curve formed by a chain that is supported on both ends at equal heights; today it is often seen in the construction of bridges. The catenary curve can be shown on the xy-plane as a chain dropping at $x = 0$ to its lowest height, $y = a$; it is given by the equation $y = \left(\dfrac{a}{2}\right)e^{\frac{x}{a}} + e^{\frac{-x}{a}}$.

It can also be expressed in terms of a hyperbolic cosine function via the equation $y = a\cosh\left(\dfrac{x}{a}\right)$. By 1695, he further enhanced the design of bridges by applying calculus to his analyses.

As the two Bernoulli brothers further grappled with applications of calculus, which was an area that was not clearly understood by many mathematicians at the time, a rivalry between them evolved. They began to criticize each other in print, and at the same time challenge each other with mathematical problems, which has moved the understanding of mathematics further along. By 1697, their relationship completely dissolved and they separated and were no longer in communication.

Figure 22.2. *Ars Conjectandi*, published 1713. (Milan, Mansutti Foundation.)

Jacob Bernoulli died in Basel, Switzerland, on August 16, 1705. It is unfortunate that he did not live to see the publication in 1713 of *Ars Conjectandi* (The Art of Conjecturing), which was a summary of most of his finest discoveries (see fig. 22.2). It included, among other topics, the theory of permutations and combinations, and the famous Bernoulli numbers, which were used to compute easily the sums of powers of any consecutive integers. Discussing them, he was reported to have said, "with the help of this table, it took me less than half of a quarter of an hour to find that the 10 powers of the first 100 numbers being added together will yield the sum 91,409,924,241,424,243,424,241,924,242,500."[2] The values of the first ten Bernoulli numbers, B_n, are given in figure 22.3, where, if n is a multiple of 4 and not equal to zero, then $B_n < 0$, and all others are positive—with the possible exception of $n = 1$, which is $\pm\frac{1}{2}$. Bernoulli numbers grew out of a long term interest in the sums of integral powers, which has fascinated mathematicians since antiquity. Curiously enough, these numbers were the subject of the first complex computer program.

The Bernoulli numbers can be defined as B_k in the formula for the series sum:

$$\sum_{k=0}^{p}\frac{B_k}{k!}\cdot\frac{p!}{(p+1-k)!}\cdot n^{p+1-k}=\frac{B_0}{0!}\cdot\frac{n^{p+1}}{p+1}+\frac{B_1}{1!}n^p+\frac{B_2}{2!}pn^{p-1}+\frac{B_3}{3!}p(p-1)n^{p-2}+\cdots+\frac{B_p}{1!}n.$$

Only the ambitious reader will want to pursue this further.

Ars Conjectandi also further introduced the subject of probability and Bernoulli's now-famous law of large numbers, which are used today when sampling statistical populations. The law of large numbers states that the frequencies of events with the same likelihood of occurrence even out over time when there are many trials or events. As the number of events increases, the actual ratio of outcomes will converge on the theoretical, or expected, ratio of outcomes.

n	B_n
0	1
1	$\pm\dfrac{1}{2}$
2	$\dfrac{1}{6}$
3	0
4	$-\dfrac{1}{30}$
5	0
6	$\dfrac{1}{42}$
7	0
8	$-\dfrac{1}{30}$
9	0
10	$\dfrac{5}{66}$

Figure 22.3. Bernoulli numbers for $n = 0$ through $n = 10$.

Bernoulli's work was notable for three reasons. First, he performed his research with only a superficial understanding of those who had come before him—he was able to read a copy of Christiaan Huygens' *Reasoning in Games of Chance*—but it's clear from his work that he had not read the letters of Pascal and Fermat, Pascal's *Treatise*, and several other texts that would have informed his research. Second, he progressed much further in the study of probability than those who came before him, despite not having access to their writings. Third, and finally, he undertook to explain not only games of chance, where the outcomes are assumed to be fair and the sole output of cards, dice, or coins; he also sought to explain human problems such as decision making. Jacob Bernoulli's works were published posthumously in a two-volume set titled *Opera Jacobi Bernoulli*, in 1744 (see fig. 22.4).

We now turn to Johann Bernoulli (see fig. 22.5), one of Jacob Bernoulli's younger brothers, who was born in Basel, Switzerland, on August 6, 1667, twelve years after Jacob's birth, and was the tenth child of Nicolaus and Margaretha Bernoulli. His father, Nicolaus Bernoulli, was a pharmacist

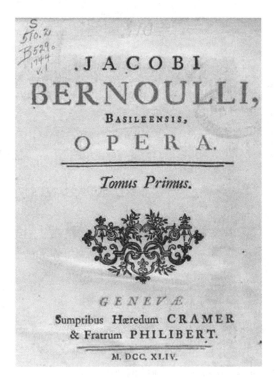

Figure 22.4. Volume 1 of *Opera Jacobi Bernoulli*.

Figure 22.5. Johann Bernoulli.
(Oil on canvas by Johann Rudolf Huber, 1740.)

and wanted his son Johann eventually to take over his business and there-fore guided Johann's university studies in that direction.

Not interested in studying business, Johann elected to study medicine, in an attempt to somewhat satisfy his parents. But this subject also did not interest him. His older brother Jacob enticed him to consider mathematics, which ended up being the right fit for him. The brothers engaged them-selves initially in the new subject of calculus (see fig. 22.6), which, as men-tioned earlier, was developed by Leibniz.

Despite his overriding interest in mathematics, he did complete his studies in medicine at the University of Basel, where he received his doc-torate in 1694. Much to his father's disappointment, he subsequently im-mersed himself in mathematics, publishing two books on differential and integral calculus. Soon thereafter, in 1694, he married Dorothea Falkner, with whom he had three boys, one of whom was Daniel, whose career we shall consider later. This was also a time when he began his position as pro-fessor of mathematics at the University of Groningen in the Netherlands. Johann seems to have been rather happy at Groningen, as evidenced by a

Figure 22.6. Jacob and Johann Bernoulli. (Engraving ca. 1870–1874.)

note he wrote on July 25, 1696, in which he stated *"Patria est, ubi bene est"* ("Where there is bread, there is my homeland") (see fig. 22.7).

Throughout his life, he has had some very famous students, such as Leonhard Euler and the French mathematician Guillaume-François-Antoine, marquis de l'Hôpital (1661–1704), who is still known today for the mathematical procedure known as l'Hôpital's rule, which is used in differential calculus. Curiously, l'Hôpital offered to pay Johann, if he would provide him with some mathematical discoveries. Interestingly, Johann Bernoulli later signed a contract with l'Hôpital, which allowed l'Hôpital to use Johann's work without proper attribution, and thus he published *Analyse des infiniment petits pour l'intelligence des lignes courbes* in 1696, which mainly consisted of the work of Johann Bernoulli, including what we know today as l'Hôpital's rule (see fig. 22.8).

The study of calculus became ever more popular and was further popularized when Johann posed the brachistochrone problem, which engaged a variety of mathematicians at the time. The problem involved taking a wire attached at two points at different heights, and placing a bead on the wire. Then letting the bead slide along the wire (assuming no friction), from the higher-height endpoint. The challenge was to determine what the shape of the curve should be in order for the bead to land at the lowest point in the

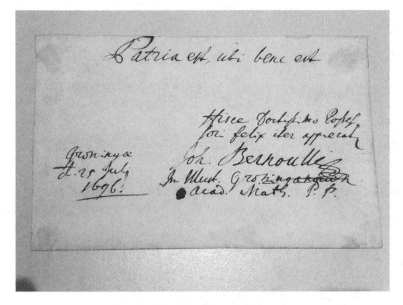

Figure 22.7. Johann Bernoulli's note.

least amount of time. Using calculus, it was determined by him that the curve is an inverted cycloid. The cycloid curve can also be generated by the path that a point on a circle travels while the circle is rolling along a straight-line path, as shown in figure 22.9.

In 1705, Johann's family urged him to return to Basel. During the journey back, he learned of Jacob's death from tuberculosis. Initially, Johann had returned to Basel so as to assume the professorial chair for Greek at the University of Basel. After the death of his brother, though, he was able to get the now-vacant position of professor of mathematics at the university of Basel, which previously had been held by his brother. It must be said that, despite the loss of Jacob, Johann was delighted to change his plans when the mathematics position became available.

In 1713, Johann got actively involved in supporting Leibniz in the discussions about who should be credited with discovering calculus. He showed that Leibniz's work was able to solve problems that Newton's fluxions could not accomplish. As a further effort to support Leibniz's position regarding the development of calculus, he published a text on integral calculus in 1742 and soon thereafter a text on differential calculus.

Apparently, Johann Bernoulli had a rather-jealous personality, which caused his previously mentioned competition and subsequent fallout with

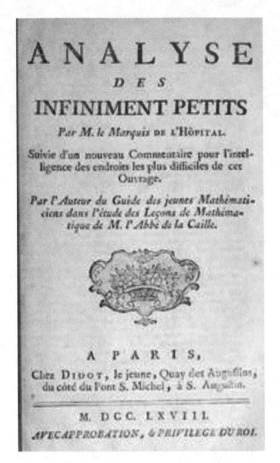

Figure 22.8. *Analyse des infiniment petits pour l'intelligence des lignes courbes*, 1696.

his brother. A similar problem disrupted his relationship with his son Daniel Bernoulli, who was also a very gifted mathematician. In 1734, Daniel wrote an important work, *Hydrodynamica*, which he published in 1738. This was about the same time that his father, Johann Bernoulli, published his work *Hydraulica*. Once again, a dispute evolved about the ownership of the material. It is believed that Johann plagiarized from his son Daniel's work.[3] This further destroyed their relationship. Johann Bernoulli died on January 1, 1748, in Basel, Switzerland, while still at odds with his son Daniel.

Johann Bernoulli's son Daniel was born on February 8, 1700, in Groningen, Netherlands. Despite his father's urging that Daniel study business, Daniel insisted on studying mathematics. He later delved into some business study but ended up studying medicine at his father's suggestion, with the understanding that his father would teach him mathematics at home.

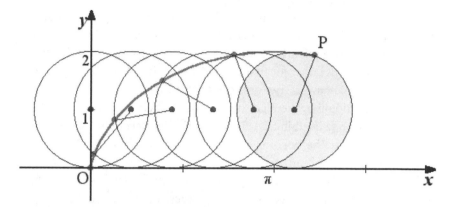

Figure 22.9. Cycloid curve formed by the path of a point
on a circle that is rolling along a straight-line path..

Figure 22.10. Daniel Bernoulli. (Portrait on
albumen paper copy mounted on cardboard, ca. 1750.)

Eventually, in 1721, Daniel earned his doctorate in anatomy and botany. As already mentioned, his father was a rather jealous person. This manifested itself once again in a scientific contest at the University of Paris, when Daniel was tied with his father for first place. As a result, Johann banned Daniel from his house. Johann carried this grudge for the rest of his life.

Daniel Bernoulli befriended the famous mathematician Leonhard Euler, who at the time lived in St. Petersburg; Daniel Bernoulli traveled there in 1724 to assume the position of professor of mathematics. He did not seem to enjoy the position, a situation made even worse by a brief illness he suffered in 1733. So he returned to the University of Basel, where he held professorial positions in medicine, metaphysics, and natural philosophy.

Daniel Bernoulli was a brilliant mathematician and scientist whose talents clearly rivaled those of his father. Eventually, though, the rivalry became counterproductive, since he won many prizes, some of which his father felt should have been awarded to the father and resultingly evicted Daniel from his home. We could ask ourselves, what might be the most important discovery that Daniel Bernoulli provided future science? In what is known today as the Bernoulli effect, Daniel proved that when a fluid flows through a region in which its speed increases, its pressure will fall. He correctly described the effect mathematically. The Bernoulli effect has many real-life applications, and it explains why aircraft wings provide lift to the plane. Essentially, what this explains is that the wing is shaped so that air flows faster over the upper part of the wing than over the lower. This results in an air pressure difference that produces lift. After a rather full life of brilliance, Daniel died on March 17, 1782, in Basel, Switzerland

Thus, we have a good overview of what is probably the most famous mathematics family, the Bernoullis. They interacted with the leading mathematicians over a century to provide many significant breakthroughs in the development of mathematics. They have contributed not only to mathematics but also to science, philosophy, business, and a general knowledge to help us better understand our environment.

CHAPTER 23

Leonhard Euler:
Swiss (1707–1783)

One often wonders from where our many mathematical symbols stem. The answer is quite simple. Perhaps one of the most prolific mathematicians of all time, the Swiss mathematician Leonhard Euler, introduced a significant number of the symbols that we frequently use in mathematics today. Although the Greek letter π was first used by the Welsh mathematician William Jones (1675–1749), it was Euler who through his many publications popularized the symbol to represent the ratio of a circle's circumference to its diameter. He also used the Greek letter Σ to represent a summation, and the letter i to represent the imaginary number $\sqrt{-1}$. He was the first to use the letter e to represent the natural logarithm, which is approximately equal to 2.71828 . . ., and that allowed him to set up the famous Euler identity $e^{i\pi} + 1 = 0$, which uses all of these symbols. Euler also introduced the concept of the function and was the first to write it as $f(x)$. We can thank Euler for the modern notation of the trigonometric functions. Even the way we today label geometric figures such as a triangle—whose vertices are marked with the letters A, B, and C, and whose notation for the sides opposite these vertices use the lowercase letters a, b, and c—stems from Euler's writings. Euler provided us with our mathematical language, but it resulted from the many volumes that he wrote during his seventy-six-year life span. We must also acknowledge the multitude of innovations that Euler introduced in mathematics. But first, let's get a brief view into his life's history.

Figure 23.1. Leonhard Euler. (Portrait by Jakob Emanuel Handmann,
pastel on paper, 1753.)

Leonhard Euler was born on April 15, 1707, in Basel, Switzerland. His father was a pastor and his mother was the daughter of a pastor, and he was one of four children. When Leonhard was a small child, his family moved to the town of Riehen, Switzerland, where he lived until age thirteen, at which time he moved back to Basel to live with his maternal grandmother. There, in 1720, the thirteen-year-old Leonhard enrolled as a student at the University of Basel; in 1723, at age seventeen, he received a master's degree in philosophy, comparing the ideas of Newton and Descartes. It was during that time that his father's friend, the famous Swiss mathematician Johann Bernoulli, gave him private lessons in mathematics and discovered his unique talents in this field. As a result, Bernoulli convinced Leonhard's father that his son should pursue a study of mathematics rather than fulfilling his father's wish of also becoming a pastor. Euler completed his dissertation on the propagation of sound in 1726 but was unsuccessful in obtaining a faculty position at the University of Basel. In 1727, Euler accepted a position in St. Petersburg in the mathematics department at the Imperial Russian Academy of Sciences. This academy was eager to recruit scholars from other European countries and was chiefly interested in research rather than in teaching students. During this time, Euler mastered the Russian language and also took on an additional job as a medic in the Russian Navy. In 1731, he was promoted to professor of physics; two years later, he headed the mathematics department as well, after the position was vacated by Daniel Bernoulli, Johann Bernoulli's son.

In 1734, Euler married Katharina Gsell, with whom he had thirteen children. Only five of those children survived childhood, and only three survived him. In 1738, apparently through a protracted fever and overwork, Euler lost sight in his right eye. This did not distract him from continuing his work as energetically as he had previously. In 1741, the instability in Russia motivated Euler to take on a faculty position at the Berlin Academy. He stayed there for the next twenty-five years and wrote over 380 articles. During his Berlin years, he also wrote two very famous books: *Introductio in analysin infinitorum*, on the topic of functions, was published in 1748; and *Institutiones calculi differentialis*, on differential calculus, was published in 1755. Later, in 1770, he completed a book titled *Institutionum calculi integralis*. These last two books provide many formulas for differentiation and integration, which compose much of our modern-day calculus course.

In 1766, Euler accepted an invitation from Catherine II to return to Russia, where soon after his arrival in St. Petersburg, a cataract formed in his remaining good eye, which left him totally blind for the rest of his life. Amazingly, this did not reduce his productivity, largely because of his uncommon memory and unusual ability to do mental calculations. He actually claimed that losing sight enabled him to have fewer distractions in his work. With his scribes, he was able to produce even more work than he had previously. Although he was not considered a teacher of mathematics, he did have a great influence on mathematics education in Russia. Euler touched so many areas of mathematics that it would take volumes to summarize all of his work. However, we will cite two results of his genius that still remain relatively popular today, even beyond mathematical circles.

First, there is a famous problem in mathematics that stems from an age-old conundrum that intrigued folks in Europe for many years. We begin with a little bit of historical background so that you will become fascinated by the problem that faced generations of Europeans.[1]

In the eighteenth century and earlier, when walking was the dominant form of local transportation, people would often count particular kinds of objects they passed. One such was bridges. Through the eighteenth century, the small Prussian city of Königsberg (today called Kaliningrad, Russia), located where the Pregel River forms two branches around an island portion of the city, provided a recreational dilemma: Could a person walk over each of the seven bridges *exactly once* in a continuous walk through the city? The residents of the city had this as a recreational challenge, particularly on Sunday afternoons. Since there were no successful attempts, the challenge continued for many years.

This problem provides a wonderful window into the field today known as networks, which is also referred to as graph theory, an extended field of geometry. This problem gives us a nice introduction into the subject. Let us begin by presenting the problem. In figure 23.2 we can see the map of the city with the seven bridges.

In figure 23.3, we indicate the island with *A*, the left bank of the river with *B*, the right bank with *C*, and the area between the two arms of the upper course with *D*. The seven bridges are called Holz, Schmiede, Honig, Hohe, Köttel, Grüne, and Krämer (see fig. 23.3). If we start at Holz and walk to Schmiede and then through Honig, through Hohe, through

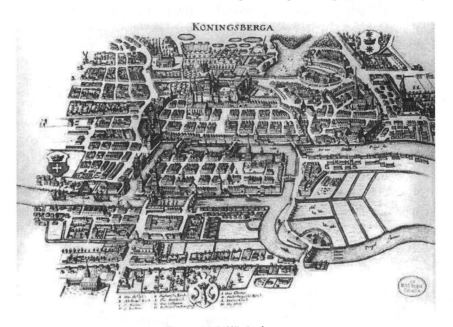

Figure 23.2. Königsberg.

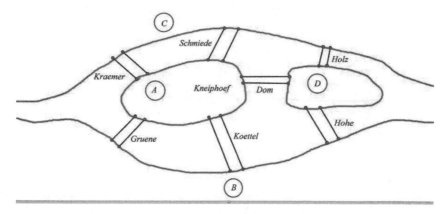

Figure 23.3. Königsberg Bridges Problem.

Köttel, through Grüne we *will* never cross Krämer. On the other hand, if we start at Krämer and walk to Honig, through Hohe, through Köttel, through Schmiede, and then through Holz, we will never travel through Grüne.

In 1735, Euler proved mathematically that this walk could not be performed. The famous Königsberg Bridges Problem, as it has become known, is a lovely application of a topological problem with networks. It is very nice to observe how mathematics—used properly—can put a practical problem to rest.

Before we embark on the problem, we ought to become familiar with some basic concepts involved. Toward that end, try to trace with a pencil each of the configurations shown in figure 23.4 without missing any part and without going over any part twice. Make sure to keep count of the number of arcs or line segments, which have an endpoint at each of the points A, B, C, D, and E.

Configurations, or networks, such as the five figures shown in figure 23.4, are made up of line segments and/or continuous arcs. The number of arcs or line segments that have an endpoint at a particular vertex is called the *degree* of the vertex.

You should notice two direct outcomes after trying to trace the networks as described above: The networks can be traced (or traversed) if they have (1) all even degree vertices or (2) exactly two odd-degree vertices. The following two statements summarize this finding:

1. There is an even number of odd-degree vertices in a connected network.

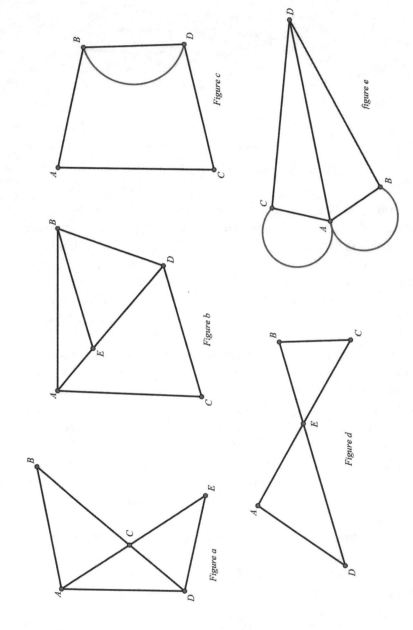

Figure a

Figure b

Figure c

Figure d

figure e

Figure 23.4. Networks.

2.   A connected network can be traversed only if it has at most two odd-degree vertices.

Let's examine each of the five networks:

Network figure 23.4a has five vertices. Vertices *B*, *C*, *E* are of even degree, and vertices *A* and *D* are of odd degree. Since figure 23.4a has exactly two odd-degree vertices as well as three even-degree vertices, it is traversable. If we start at *A* then go down to *D*, across to *E*, back up to *A*, across to *B*, and down to *D*, we have chosen a desired route.

Network figure 23.4b has five vertices. Vertex *C* is the only even-degree vertex. Vertices *A*, *B*, *E*, and *D* are all of odd degree. Consequently, since the network has more than two odd-degree vertices, it is not traversable.

Network figure 23.4c is traversable because it has two even-degree vertices and exactly two odd-degree vertices.

Network figure 23.4d has five even-degree vertices and, therefore, can be traversed.

Network figure 23.4e has four odd-degree vertices and *cannot* be traversed.

The Königsberg Bridge Problem is the same problem as the one posed in figure 23.4e. Let's take a look at both figure 23.4e and figure 23.3 and note the similarity. There are seven bridges in figure 23.3, and there are seven lines in figure 23.4e. In figure 23.4e, each vertex is of odd degree. In figure 23.3, if we start at *D*, we have three choices: we could go to Hohe, Honig, or Holz. If we start at *D* in figure 23.4e, we have three line paths to choose from. In both figures, if we are at *C*, we have either three bridges or three lines we could go on. A similar situation exists for locations *A* and *B* in figure 23.3 and vertices *A* and *B* in figure 23.4e. We can see that this network cannot be traversed.

By reducing the bridges and islands to a network problem, we can easily solve it. This is a clever tactic to solve problems in mathematics. You might want to try to find a group of local bridges in your region to create a similar challenge, and then see if the walk is traversable. This problem and its network application is an excellent introduction into the field of topology. Getting actively involved with the various experiments we described above would ensure a genuine understanding of an aspect of mathematics to which most are not otherwise exposed.

Among the enormous number of contributions that Euler made to the field of mathematics, there is one that can appeal very nicely to the novice. Euler established that for any convex polyhedron, the relationship between the number of vertices (V), edges (E), and faces (F) satisfies the following equation: $V + F = E + 2$, which is rightfully known as the Euler formula. It might be fun to verify this formula with any convex polyhedron you may have available. You might begin with the five regular polyhedra shown in figure 23.5 by counting the number of vertices, faces and edges and see that in each case Euler's formula holds true.

It should be mentioned that Euler was the most prolific mathematician and scientist, yes, but he was also heavily involved in areas beyond pure mathematics, such as cartography, physics, and astronomy, just to name a few. We still remember Euler today as having contributed to mathematics more volumes of work than any other mathematician in history has. A large number of mathematical objects and topics are named in honor of Leonhard Euler. In fact, he made so many pioneering contributions to several branches of mathematics that some theorems were deliberately attributed to the first mathematician to have discovered them after Euler, in order to avoid naming everything after Euler.

On September 18, 1783, during a discussion with his two assistants on the topic of planetary motion, and more specifically about the recently discovered planet Uranus, Euler collapsed from a brain hemorrhage and died a few hours later. To further exemplify Euler's productivity, the St. Petersburg Academy continued to publish Euler's findings for another 50 years beyond his death.

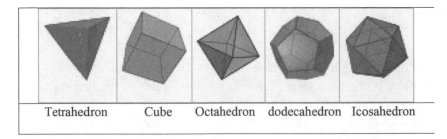

| Tetrahedron | Cube | Octahedron | dodecahedron | Icosahedron |

Figure 23.5.

CHAPTER 24

⌒

Maria Gaetana Agnesi: Italian (1718–1799)

Looking at the history of mathematics, we can see readily that it is overloaded with male mathematicians. This prompts us to wonder, who was the first female mathematician in the Western world to receive an international reputation? Most people would consider Maria Gaetana Agnesi to fit that role. She was born on May 16, 1718, in Milan, which was then a part of the Austro-Hungarian Empire, and is now a city in Italy. She grew up in a wealthy household as the oldest child of a prosperous silk merchant, who fathered twenty-one children. The first signs of her prodigy status appeared at age five, when she spoke Italian and French. By age nine, she further demonstrated her talent by mastering several modern languages, as well as Latin, Greek, and Hebrew. Within the next few years, she mastered mathematics.

Recognizing her talents, her father would invite friends and have her perform by displaying her vast intelligence. Interestingly, in 1738, when she was twenty years old, a series of essays were published, *Propositiones philosophicae* (Propositions of Philosophy), which were based on her presentations at these forums. In 1748, she published *Instituzioni analitiche ad uso della gioventù italiana* (Analytical Institutions for the Use of Italian Youth), which encompassed two large volumes, wherein she presented her treatment of algebra and both integral and differential calculus (see fig. 24.2). This work was well received by mathematicians all over Europe. A committee of the renowned Académie des Sciences in Paris reported on *Instituzioni analitiche ad uso della gioventù italiana*:

Figure 24.1. Maria Gaetana Agnesi. (Engraving by G. A. Sasso
after G. B. Bosio, Library of Congress.)

It took much skill and sagacity to reduce, as the author has done, to
almost uniform methods these discoveries scattered among the works
of modern mathematicians and often presented by methods very different from each other. Order, clarity and precision reign in all parts
of this work. . . . We regard it as the most complete and best made
treatise.[1]

One form of her fame came from a cubic curve known in Italian as *ver-
siera*, which, over the years, has become confused with the word for "witch,"
namely, *versicra*. Ultimately, this resulted in its English name, the Witch of
Agnesi.

The bell-shaped curve, which we refer to as the Witch of Agnesi, can
be constructed as follows: Start with a circle of diameter *a*, centered at the
point $(0, \frac{a}{2})$ on the *y*-axis, as shown in figure 24.3. Then select a point *A*

INSTITUZIONI
ANALITICHE
AD USO
DELLA GIOVENTU' ITALIANA
DI D.ᴺᴬ MARIA GAETANA
AGNESI
MILANESE
Dell' Accademia delle Scienze di Bologna.
TOMO I.

IN MILANO, MDCCXLVIII.
NELLA REGIA-DUCAL CORTE.
CON LICENZA DE' SUPERIORI.

Figure 24.2.

on the line $y = a$, and draw the line segment AO, with its intersection with circle O at point B. Let point P be the point where the vertical line through A intersects the horizontal line through B. The curve , called "Witch of Agnesi," is then traced by point P as A moves along the line $y = a$.

The curve has the equation $yx^2 = a^2(a - y)$, or $y = \dfrac{a^3}{x^2 + a^2}$.
Figure 24.4 shows the curve in its original rendering.

This entire work so impressed Pope Benedict XIV that, in 1750, he appointed Agnesi to the position of professor of mathematics at the University of Bologna. Soon thereafter, she increasingly gravitated to religious studies and never actually visited Bologna again. After her father died in 1752, Maria Gaetana Agnesi completely dedicated herself to religious studies and charitable work. She died on January 9, 1799, in one of the charity poorhouses that she directed.

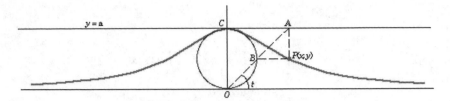

Figure 24.3.

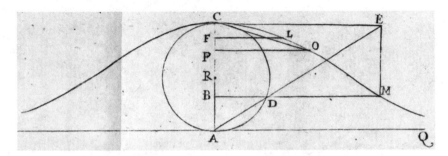

Figure 24.4. "Witch of Agnesi." (Maria Gaetana Agnesi, *Instituzioni analitiche ad uso della gioventù italiana*, 1748.)

CHAPTER 25

Pierre Simon Laplace:
French (1749–1827)

In the early 1980s, the American neuroscientist Benjamin Libet (1916–2007) conducted a now-famous series of experiments questioning the existence of "free will."[1] In these experiments, participants were placed in front of a rapidly moving clock and instructed to carry out some small motor activity (such as flexing a finger or clenching a fist) whenever they felt like it, and note the position of the hand of the clock at the instant they were aware of the intention to act. The experiment showed that the decision to act occurs approximately 200 milliseconds prior to the movement. In addition, the researchers monitored the subjects' brain activity during the whole experiment, using an electroencephalogram (EEG). Surprisingly, the EEG signal indicated brain activity related to the resulting movement more than 500 milliseconds in advance of the action, and, thus, more than 300 milliseconds before the subjects reported their first awareness of their desire to act. These findings seemed to suggest that whenever we make an apparently conscious decision to do something, we are actually just executing what our brain has already decided. Does this mean that free will is just an illusion? Is everything we do and everything we experience controlled by a subconscious program running in our brain, creating our consciousness and just making us believe that we are free to choose our actions? These questions are still not answered, and they are the subject of ongoing scientific debates. We have not yet understood how consciousness works, that is, how physiological processes such as ions flowing across nerve membranes cause us to have experiences. As long as this has not been clarified (and it is an extremely complex problem to solve), we cannot expect a definite answer to the question of existence or non-existence of conscious free will. It was already recognized by ancient Greek philosophers that free will is in conflict with the philosophical idea that all events are determined by previously existing causes, a philosophical theory known as determinism. If every event has a unique cause, then complete knowledge of the present state of the universe would, at least in principle, enable us to predict its future. But if the future is already determined by the present, then there is no room for free will. Although the concept of determinism goes back to philosophers in ancient Greece, the notion is nowadays strongly associated with Newtonian mechanics, which is the mathematical description of the motion of bodies under the influence of forces, based on Newton's famous laws of motion. Newton's three laws of motion were first published in 1687 in his opus magnum *Philosophiæ Naturalis Principia Mathematica* (Mathematical Principles of Natural Philosophy), commonly known as the *Principia*. They are fundamental laws of physics, describing the motion of colliding billiard balls just as well as the motion of planets in the solar system. Newtonian mechanics is deterministic in the sense that, if one knows the positions and velocities of all matter particles at some instant in time, one can, at least theoretically, calculate their future positions and velocities. Newton's equations of motion provided the precise mathematical relationship between cause and effect, thereby establishing a solid scientific basis for the philosophical idea of determinism. The French mathematician Pierre Simon de Laplace (1749–1827) was among the first who recognized the philosophical implications of a mechanistic view of the universe. The following excerpt from his "Philosophical Essay on Probabilities," published in 1814, is the first known articulation of scientific or causal determinism:

> We may regard the present state of the universe as the effect of its past and the cause of its future. An intellect which at a certain moment would know all forces that set nature in motion, and all positions of all items of which nature is composed, if this intellect were also vast enough to submit these data to analysis, it would embrace in a single formula the movements of the greatest bodies of the universe and those of the tiniest atom; for such an intellect nothing would be uncertain and the future just like the past would be present before its eyes.[2]

This hypothetical intellect became later famously known as "Laplace's demon," although Laplace did not use the word "demon." But Pierre Simon de Laplace is not only remembered as a pioneer of scientific determinism, he also made fundamental contributions to probability theory, statistics, and to the theory of differential equations. Most important, he introduced differential calculus into physics and astronomy, thereby substantially extending the work of Newton and other predecessors. His most important treatise, *Traité de mécanique céleste* (Treatise of Celestial Mechanics), was a seminal contribution to mathematical physics; it became a standard text in astronomy and remained state-of-the-art for more than one hundred years. We will examine a few of his scientific achievements in more detail as we provide a short of summary of his life.

Pierre Simon Laplace was born on March 23, 1749, in Beaumont-en-Auge, Normandy, France.[3] His father, Pierre Laplace, was a cider merchant, and his mother, Marie-Anne Sochon, came from a relatively prosperous

Figure 25.1. Pierre Simon Laplace. (Painting by Jean-Baptiste Paulin Guérin, oil on canvas, 1838.)

farming family. There is no evidence of higher academic education in his family. Pierre Simon attended the Benedictine priory school in Beaumont-en-Auge between the ages seven and sixteen. His father wanted him to become a priest, and so he sent him to the University of Caen to study theology. However, Laplace soon discovered that he was much more attracted to mathematics than to theology. This shift of interests was partly provoked by two inspiring mathematics teachers at Caen, Christophe Gadbled and Pierre Le Canu. Fortunately, they realized Laplace's talent for mathematics and began to mentor him. While still a student at Caen, Laplace wrote his first mathematical paper and sent it to the renowned Italian mathematician Joseph-Louis Lagrange (1736–1813), who would later publish it in his journal, *Miscellanea Taurinensia*. Laplace was now determined to become a professional mathematician, and Le Canu wrote a letter of introduction for him to Jean-Baptiste le Rond d'Alembert (1717–1783), one of the leading mathematicians in Paris. Without taking his degree, Laplace traveled to Paris to introduce himself to d'Alembert. Descendants of d'Alembert reported that he was initially very reserved when he received Laplace in Paris.[4] To get rid of him, he handed him a thick book of advanced mathematics, and told him to come back after he had read it. Not expecting to see Laplace again in the near future, d'Alembert was perplexed when Laplace knocked on his door only a few days later. He didn't believe that Laplace had read the whole book, let alone that he had understood it. Visibly annoyed, he started to ask him mathematical questions, and soon had to acknowledge that Laplace had indeed understood everything contained in the book. His initially reluctant attitude disintegrated, and the more problems d'Alembert posed to Laplace, the more impressed he became with Laplace's mathematical brilliance. He welcomed Laplace as his student and found a position for him as a mathematics professor at the École Militaire, a military academy in Paris. Financially secure and with not too many teaching obligations, Laplace delved into research and soon produced substantial mathematical results. After having published several high-quality papers, he applied for a position at the Academy of Sciences in Paris, the most distinguished scientific institution in France at that time. However, still in his early twenties, he was passed over in favor of older, but less prolific, mathematicians. Laplace, who was never modest about his achievements, became very angry about the decision he felt was unjust. Yet he didn't have to wait too long. In 1773, after two unsuccessful attempts, he became an adjunct member of the Academy of Sciences (and was promoted to a senior position in 1785). With an uninterrupted flow of important contributions to mathematics during the 1770s, Laplace's reputation as a mathematician steadily increased. In those years, he focused his interests and developed both his style as a mathematician and his philosophical viewpoints. His main fields of research gradually took shape, namely, the application of differential calculus to the motion of astronomical objects and the theory of probability. Understanding and calculating the motion of the planets in the solar system had been a big mathematical challenge throughout the history of science.

In his 1687 publication *Philosophiae Naturalis Principia Mathematica*, Newton showed that Johannes Kepler's laws for planetary motion follow from Newton's own three fundamental laws of mechanics, combined with his law of universal gravitation. The latter stated that two massive bodies attract each other with a force directly proportional to the product of their masses and inversely proportional to the square of the distance between their centers. Kepler had found his laws for the motion of planets empirically, by carefully studying observational data, but Newton "proved Kepler's laws" in deriving them from his fundamental laws of mechanics. This was a great success for Newton's theory; however, to derive Kepler's laws from his equations of motion, Newton had to consider a single planet orbiting around the sun; in astronomy, this is called the "two-body problem." But as soon as three or more bodies are involved, the equations describing their motion become much more complicated. In fact, the "three-body problem" is already so complicated that it can only be solved in an approximate sense. The planets in our solar system do not exactly behave according to Kepler's laws; the gravitational pull they exert on each other makes their orbits deviate slightly from perfect Kepler orbits. Even the smallest planet has a tiny effect on the motion of all other planets, and, over long periods of time, these tiny disturbances may accumulate to large deviations. Describing the whole solar system in the framework of Newtonian mechanics leads to an extremely complicated system of equations, since the gravitational force acting on a planet depends not only on its distance from the sun but also on the current positions of all other planets. Newton had recognized the mathematical difficulties in his attempts, and he doubted that this complex system of equations could be solved at all. His conclusion was that periodic divine intervention was necessary for the solar system to remain stable; otherwise, the small disturbances of a planet's orbit caused by the presence of other planets could add up over time and finally kick the planet out of the solar system. Observational data gathered in the early eighteenth century indicated that Jupiter's orbit was slowly shrinking, while Saturn's orbit was expanding. Explaining this apparent instability was a big open problem of astronomy; even Leonhard Euler and Joseph-Louis Lagrange had unsuccessfully tried to solve it. Laplace carried out a more refined mathematical analysis of the problem, incorporating effects that Euler and Lagrange had omitted, and his calculations turned out to be in perfect agreement with the observational data. His analysis revealed that the special ratio between the orbital periods of Jupiter and Saturn is responsible for the anomalies in their motions. (Two periods of Saturn's orbit around the sun are almost exactly equal to one period of Jupiter's orbit.) Having solved this longstanding problem, Laplace aimed at a theoretical description of the whole solar system. His scientific goal was to: "bring theory to coincide so closely with observation that empirical equations should no longer find a place in astronomical tables."[5]

In his major work, *Celestial Mechanics*, published in five volumes between 1799 and 1825, Laplace brought the methods of calculus into classical mechanics, which previously had mainly been studied geometrically. The powerful machinery of calculus enabled Laplace to tackle problems Newton and other predecessors had considered too complicated for a mathematical analysis. Laplace developed a complete mathematical framework for calculating the motions of the planets and their satellites, including the tidal motion and the effects of tidal forces on the shape of planets. In particular, he was able to show stability of the solar system[6] without having to postulate any divine intervention, as Newton did. There is a famous anecdote regarding a conversation between Laplace and Napoleon Bonaparte: When Laplace presented his work to Napoleon, he was congratulated but asked why he had nowhere mentioned God in his book. Laplace's blunt and now-famous answer was: "I had no need of that hypothesis."[7]

Because Laplace was rather opportunistic with his political opinions, he escaped imprisonment during the French Revolution and was even appointed as minister of the interior by Napoleon, as a placeholder for Napoleon's brother. That is probably the main reason why Laplace's political career lasted only six weeks; however, Napoleon later wrote in his memoirs:

> Laplace did not consider any question from the right angle: he sought subtleties everywhere, conceived only problems, and finally carried the spirit of "infinitesimals" into the administration.[8]

In 1812, Laplace published his *Théorie analytique des probabilités* (Analytic Theory of Probability). The mathematical theory of probability has its origin in a correspondence between Blaise Pascal (1623–1662) and Pierre de Fermat (1607–1665). However, it was Laplace who developed a complete mathematical theory around the exemplary problems discussed by his predecessors. His views on the subject are best expressed in the following quote:

> The theory of probabilities is at bottom nothing but common sense reduced to calculus; it enables us to appreciate with exactness that which accurate minds feel with a sort of instinct for which of times they are unable to account.[9]

Laplace's book on probability theory was one of the most influential books in mathematics. He laid the foundation of the subject and formulated its basic principles, in addition to establishing a theoretical framework. He considered not only many practical applications, such as mortality, life expectancy, the length of marriages, but also triangulation methods in surveying, and other problems of geodesy, the science of measuring and understanding Earth's geometric shape and its gravitational field. Among the many results he proved, we shall just mention one of the mathematical concepts that bear his name today: Laplace's rule of succession. This rule can be seen as follows: Assume you are conducting an experiment that can result in only a success or a failure, and you repeat it independently n times. If you get k successes, what would be your estimate for the probability that the outcome of the next trial will be a success? A reasonable, and quite natural, answer would be $P = \frac{k}{n}$, but Laplace showed that in some situations, the ratio $\frac{k+1}{n+2}$ gives a better estimate. For a large n, the difference between the two ratios is negligible, but if n is small, Laplace's formula is more useful, as the following example illustrates:

If you toss a coin three times and it always comes up heads (which we here will consider success), then the formula $P = \frac{k}{n} = \frac{3}{3} = 1$ would suggest that the probability for the coin to show heads after the next toss is 100 percent, which is, of course, nonsense. Similarly, if the coin always came up tails in the first three trials, the formula $P = \frac{k}{n} = \frac{0}{3} = 0$ would suggest that the probability for the coin to come up heads in the next toss is zero. Laplace's formula avoids these wrong conclusions by leaving room for the possibility that the fourth toss might produce a different result.

In his later years, Laplace tried to extend his mathematical methods to other areas of physics. He developed a theory of light and a theory of heat (both of which proved to be wrong), and he continued to publish papers well into his seventies. Laplace died in Paris in 1827, but his name is still very present in mathematics and physics thanks to the mathematical objects and notions carrying his name. In particular, we have the well-known Laplace's equation, and the Laplace transform. (The Laplace equation is a differential equation that is important in several branches of physics; for example, the gravitational field of the Earth must satisfy Laplace's equation everywhere outside the Earth. In higher mathematics, a simple definition for the Laplace transform would be that it takes a function of a real variable and transforms it into a function of a complex variable.) Last but not least, there is Laplace's demon, who is still wandering around in philosophical discussions stimulated by new findings in neuroscience.

~

CHAPTER 26

Lorenzo Mascheroni:
Italian (1750–1800)

Unfortunately, it is not all too uncommon in the history of mathematics that a mathematical discovery is named inappropriately for a mathematician who did not initially discover a unique relationship. This is the case for the Italian mathematician Lorenzo Mascheroni, who is known today for having been the first to prove that all geometric constructions that are possible using a straight edge and a pair of compasses can be done with the compasses alone. This was presented in his book *Geometria del compasso* in 1797 (see fig. 26.1).

Before we discuss this theorem in greater detail, we should acknowledge the fact that *unbeknownst* to Mascheroni, this relationship was previously proved by the Danish mathematician Georg Mohr (1640–1697) in 1672, in a very obscure book titled *Euclides danicus*, which was rediscovered in 1928. Georg Mohr was born in Copenhagen in 1640, and in 1662 he traveled to the Netherlands to study mathematics under Christiaan Huygens. The following year, he published a second book, titled *Compendium Euclidis Curiosi*. During his time away from Copenhagen, he spent time reading the work of some of the famous mathematicians of his day in France, England, and Germany. Unfortunately, his claim to fame has been limited to Denmark, which has named its mathematics competition after him.

LA .GEOMETRIA

DEL

COMPASSO

DI

LORENZO MASCHERONI.

P A V I A anno V della Repubblica Francese.

Presso gli Eredi di Pietro Galeazzi
(1797)

Figure 26.1.

The brilliant discovery that all geometric constructions that are possible to create with a straight edge and a pair of compasses can be created with the compasses alone is actually named after Mascheroni. He came upon a proof of this theorem that is different from Georg Mohr's proof; and, since we refer to this technique as Mascheroni constructions, we will focus on Mascheroni's life here.

Lorenzo Mascheroni was born in Bergamo, Lombardi, Italy, in 1750, to a wealthy family that motivated him to become a priest, and he was ordained in 1767. Soon thereafter, he taught rhetoric, and, later on, in 1778, he taught mathematics and physics at the seminary in Bergamo. This led him to become a professor of mathematics at the University of Pavia, where in 1789 he became the rector of the university, a position he held for the next four years.

Figure 26.2. Lorenzo Mascheroni, ca. 1790.

Mascheroni so admired Napoleon Bonaparte that he dedicated his book *Geometria del compasso* to him in 1797. He also received quite a few tributes for his work, largely in geometry, by being elected to the Academy of Padua, the Italian Society of Science, and the Academy of Mantua. In 1795, the metric system was introduced in Europe, and Mascheroni was appointed to travel to Paris and study the new system so that he could report back to the government in Milan. He published a report in 1798, but he was confined to Paris as a result of the unrest stemming from Napoleon's wars affects throughout Europe. From the weather there Mascheroni caught a normal cold, which then brought on further complications leading to a fatal viral infection, resulting in Mascheroni's death on July 14, 1800.

Let us now take a careful look at what a fantastic discovery Lorenzo Mascheroni made regarding those constructions that still bear his name: Mascheroni constructions, despite the fact that Mohr seems to have first made this discovery—albeit with other techniques, and, as was already mentioned, without Mascheroni's knowledge. We will look at Mascheroni's work in some detail because the results are truly counterintuitive.

Before we actually demonstrate that the compasses can replace the unmarked straightedge to construct a straight line, we will begin by showing a few constructions that normally would involve the unmarked straightedge but here will be done with compasses alone.

In order to make our discussions more concise, we will use a shorthand method for referring to a circle, or the arc of a circle, in the following way: A circle whose center is point P and has a radius length AB will be referred to with (P, AB).[1] Also, we know that any two points determine a line, so we will refer to a line using any two of its points; for example, the line containing the points A and B will be referred to as simply AB.

We will begin with a critical construction that would be necessary to demonstrate how Mascheroni constructions can manifest themselves. Here we will attempt to find a point E on the line AB so that $AE = 2(AB)$.

Follow along as we use figure 26.3, where we begin by considering the line segment AB. We then draw arc (B, AB). We let arc (A, AB) intersect arc (B, AB) at point C. Then let arc (C, AB) intersect arc (B, AB) at point D. Further, we let arc (D, AB) intersect arc (B, AB) at point E. We can now

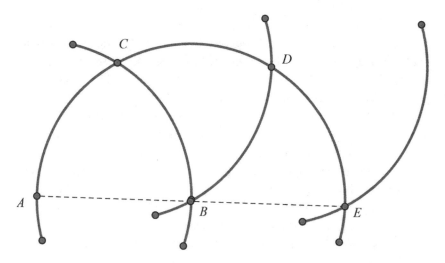

Figure 26.3.

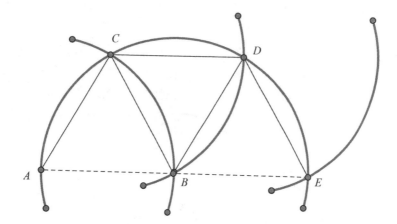

Figure 26.4.

notice that $AB = BE$, or that $AE = 2(AB)$, which we set out to construct. Some may ask, how do we know that point E, in fact, lies on line AB? If we look at triangles ABC, CBD, and DBE (fig. 26.4), we should realize that they are equilateral triangles. Therefore, the angles ABC, CBD, and DBE are each 60°, which then form a straight line, ABE.

Using the technique above, we can construct a line segment whose measure is n times the measure of any given line segment, where $n = 1, 2, 3, 4, \ldots$. We can show this in figure 26.5 by continuing the doubling of line segment AB. This will allow us to create line segments that are three times, four times, five times, and so on, as long as segment AB.

As shown in figure 26.5, we will have segments multiple times as long as segment AB. We do this as follows. We draw (E, AB) to intersect (D, AB) at point F; then we draw (F, AB) to intersect (E, AB) at point G; then (G, AB) to intersect (F, AB) at point H; then (H, AB) to intersect (G, AB) at point I; then (I, AB) to intersect (H, AB) at point J; and then (J, AB) to interest (I, AB) at point K. This process can then continue indefinitely. Notice also how we were able to place many points on the line AB, which is one of our considerations for creating the line AB with countless many points.

Now that we have shown that we can construct a line segment that is a multiple length of any given line segment along that line by adding countless many points, let us now try to find segments that are a fraction of a given segment, or say, $\frac{1}{n}$th the measure of a given line segment.

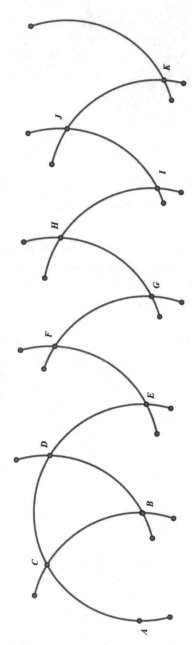

Figure 26.5.

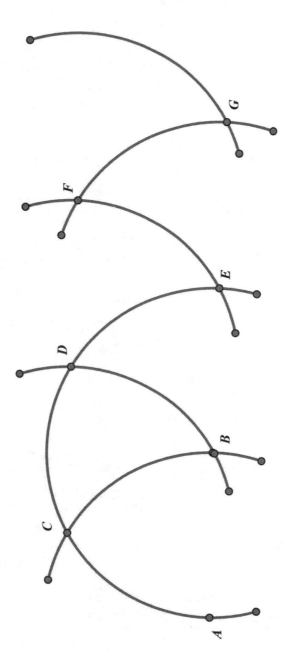

Figure 26.6.

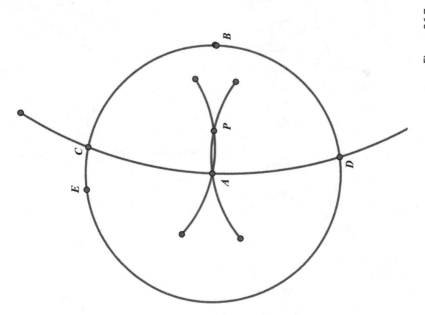

Figure 26.7.

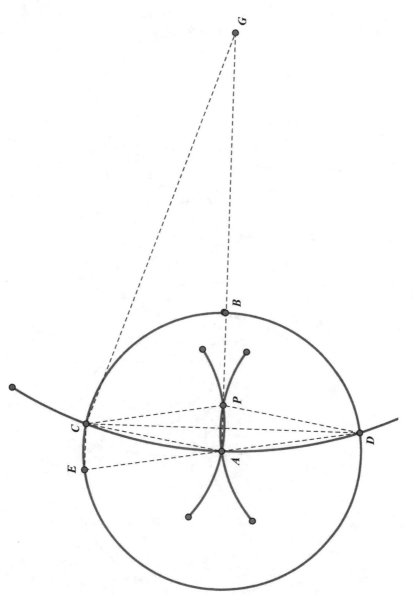

Figure 26.8.

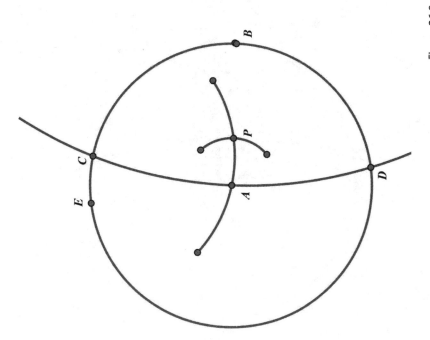

Figure 26.9.

We will begin by drawing a line segment AG, which will be three times the length of AB using the method above (see fig. 26.6).

To make matters a little clearer, we will just copy the line segment ABG, where $AB = \frac{1}{3} AG$ as shown in figure 26.6, and begin our construction that would be one-third the length of AB. We begin by drawing the circle (A, AB). Next, we draw arc (G, GA) to intersect the circle (A, AB) at points C and D, as shown in figure 26.7. The intersection-point P of arcs (C, CA) and (D, DA) is a trisection point of the segment AB, or in other words, $AP = \frac{1}{3} AB$. To find the other trisection point of AB, we merely use the process mentioned above for duplicating a line segment, in this case duplicating AP.

To better explain why this procedure works, we refer to figure 26.8, where we are adding some lines merely to explain the justification for this construction. We must first show that the point P actually lies on line ABG. The points A, P, and G lie on the perpendicular bisector of line segment CD, and, therefore, they are collinear. The two isosceles triangles CGA and PAC are similar, since they share a common base angle, namely, angle CAP. Therefore, $\frac{AP}{AC} = \frac{AC}{AG}$. However, since $AC = AB$, we have $\frac{AP}{AB} = \frac{AB}{AG}$. Since we know that $\frac{AB}{AG} = \frac{1}{3}$, we have $\frac{AP}{AB} = \frac{1}{3}$, or $AP = \frac{1}{3} AB$.

There is an alternate method for doing this construction, that is, for locating the point P: We use the first Mascheroni construction to find the point E diametrically opposite point D. Or to put it another way, DAE is the diameter of circle (A, AB). Since in figure 26.8, the quadrilateral $ECPA$ is a parallelogram, $EC = AP$. Therefore, we can find the point P by locating the intersection of arc (A, EC) and arc (C, CA). We show this in figure 26.9.

In order to justify Lorenzo Mascheroni's statement that all constructions possible with the usual geometric construction tools—the unmarked straightedge and a pair of compasses—can be done by using only compasses, as we have shown in the earlier constructions, we need not necessarily show that every imaginable construction can be done this way. Rather, we need to show only that the five following constructions are possible with compasses alone; with these five at our disposal, we are able to do all the geometric constructions that are typically created with the usual tools: a straightedge and compasses. The following five fundamental constructions are those upon which all other constructions are dependent. That is, any construction using both straightedge and compasses is merely a finite number of successions of these constructions:

1. Draw a line through two given points.
2. Draw a circle with a given center and a given radius.
3. Locate the points of intersection of two given circles.
4. Locate the points of intersection of a straight line (given by two points) and a given circle.
5. Locate the point of intersection of two straight lines (each of which is given by two points).

Although we cannot actually draw a line through the two given points, we can place as many points as we wish on the line and—perhaps working for an infinitely long time—all the points between these two points will eventually appear on that line. That would essentially satisfy the first condition listed above. The second and third constructions listed above, clearly, need no further discussion, since they are done by compasses alone. To locate the point of intersection of a straight line given by two points, say, A and B, and a given circle, (O, r), we will need to consider two cases: one for which the center of the circle is not on the given line, and one for which the center of the circle does lie on the given line.

First, we consider the case for which the center of the circle does not lie on the given line. Here we have circle (O, r) and the straight line AB, as shown in figure 26.10 (a dashed line is there merely to help us see the line AB, which was determined by only two points, A and B).

We need to find point Q, which is the point of intersection of the arcs (B, BO) and (A, AO). We then draw the circle (Q, r). The points of intersection of the circles (Q, r) and (O, r) are the required points of intersection of line AB and the circle (O, r).

This can be justified in the following way. Point Q was chosen so as to make AB the perpendicular bisector OQ. By drawing circle (Q, r) congruent to an intersecting circle (O, r), the common chord PR is also the perpendicular bisector of OQ.

The second case to consider is when the center of the given circle lies on the given line. Here, we will consider the circle (O, r) and the straight-line AB, which is shown in figure 26.11.

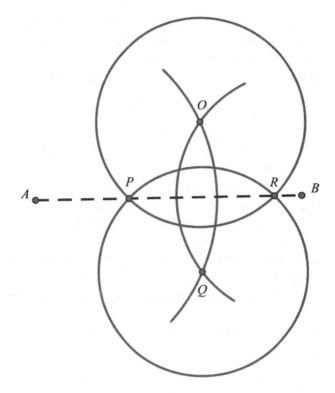

Figure 26.10

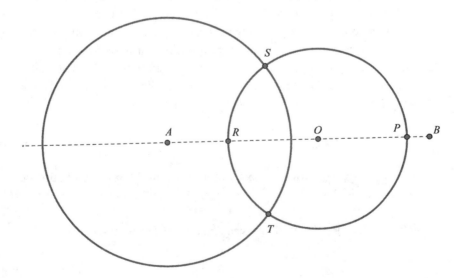

Figure 26.11

In figure 26.11, we draw circle (A, x), where radius length x is large enough to intersect circle (O, r) in two points, S and T. The midpoints of the major and minor arcs of ST, are P and R. This becomes a bit more complicated and will be shown in the following manner.

In the interest of completing the above argument, we will now focus our attention on bisecting a given arc ST. To begin our construction (see fig. 26.12) we will let $OS = OT = r$, where O is the center of the circle of which ST is an arc. We will let the distance between S and T be equal to d, and then draw the circle (O, d). We then draw the circles (S, SO) and (T, TO), which will intersect the circle (O, d) at points M and N, respectively. Next, we draw arcs (M, MT) and (N, NS); each will meet at point K. By drawing arcs (M, OK) and (N, OK), we find that their points of intersection, C and D, are the desired midpoints of the arcs ST.

In order to demonstrate why this construction does what is purported to have been done, namely to find the midpoints of arc ST, we will draw some auxiliary lines to help explain the construction as shown in figure 26.13.

Let's first look at quadrilaterals $SONT$ and $TOMS$. These quadrilaterals are parallelograms since both pairs of opposite sides are congruent. This allows us to conclude that the points M, O, and N are collinear. Since $CN = CM$, and $KN = KM$, we then can conclude that KC and MN are perpendicular at O. We can also conclude that $CO \perp ST$. Therefore, CO bisects the segment ST, and consequently the arc ST. Our remaining task is merely to show that the point C lies on circle (O, r), or to show that $CO = r$.

In order to do this, we will rely on a useful theorem in geometry that states that the sum of the squares of the measures of the sides of a parallelogram equals the sum of the squares of the measures of the diagonals.[2] Applying this to parallelogram $SONT$, we get the following: $(SN)^2 + (TO)^2 = 2(SO)^2 + 2(ST)^2$ or $(SN)^2 + r^2 = 2r^2 + 2d^2$ or, which gives us

$$(SN)^2 = r^2 + 2d^2 \tag{I}$$

By applying the Pythagorean theorem to right $\triangle KON$, we obtain the following:

$(KN)^2 = (NO)^2 + (KO)^2$. However, since $KN = SN$, we have:

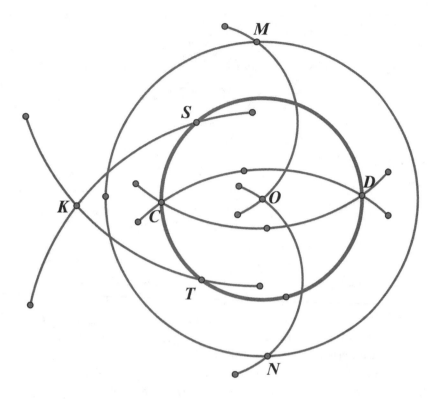

Figure 26.12

$$(SN)^2 = (NO)^2 + (KO)^2 = d^2 + (KO)^2 \tag{II}$$

Combining equations (I) and (II), we have: $r^2 + 2d^2 = d^2 + (KO)^2$, or $r^2 + d^2 = (KO)^2$.

We are now approaching the conclusion by considering right triangle CON where once again applying the Pythagorean theorem we get: $(CO)^2 + (NO)^2 = (CN)^2$ or $(CO)^2 = (CN)^2 - (NO)^2$. We know that (M, OK) and (N, OK) intersect at point C, and that CN is the radius of these two circles. Therefore, $CN = OK$. With appropriate replacements in the above equation, we get: $(CO)^2 = (KO)^2 - d^2 = r^2 + d^2 - d^2 = r^2$. Therefore, we have shown that $CO = r$, which is what we set out to demonstrate.

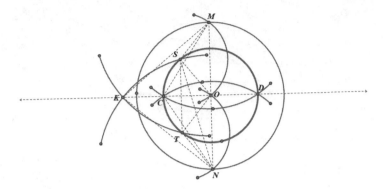

Figure 26.13

To complete our justification of the Mascheroni constructions, we need to demonstrate that the fifth construction on our original list (above) can be done with compasses alone. In other words, we now want to show that we can find the point of intersection of two straight lines, *AB* and *CD* with only a pair of compasses (see fig. 26.14). Although there are quite a few arcs to be drawn to do this construction, just follow along step-by-step—perhaps making your own drawing as you go—and the result will be rewarding.

To begin the construction, we will let arcs (*C*, *CB*) and (*D*, *DB*) meet at point *E*. Then we will let the arcs (*A*, *AE*) and (*B*, *BE*) meet at point *F*. Next, we will draw the arcs (*E*, *EB*) and (*F*, *FB*) which will meet at point *G*. Continuing with the construction we will have arcs (*B*, *BE*) and (*G*, *GB*) meet at point *H*. Finally, we will have arcs (*E*, *EB*) and (*H*, *HB*) meet at point *I*. The point we seek, namely, the intersection of the two straight lines *AB* and *CD*, is the point *M*, which is the point of intersection of the arcs (*H*, *HB*) and (*I*, *IG*).

Now comes the task of justifying that this construction does what it purports to do. Once again you will need some auxiliary lines as you will see in figure 26.15. Keep in mind that we must show that point *M* is on both line *AB* and line *CD*.

You will notice in figure 26.15 that $EI = EB = BH = HI$, since they are radii of equal circles. Similarly, $IM = IG$. We can then conclude that the arcs

IM and *IG* are congruent. The inscribed angle *IBM* has one half the measure of its intercepted arc *IM*; similarly,

$$\angle IBG = \frac{1}{2} \text{ arc } IG \,.$$

Therefore, we can conclude that $\angle IBM = \angle IBG$. This also allows us to establish that the point *M* is on line *BG*. Furthermore, we know that lines *AB* and *BG* are each perpendicular bisectors of *EF*. Again, this allows us to establish that point *M* must lie on *AB*. We now need to show that *M* also lies on line *CD*. We can easily show that triangle *BGH* and triangle *BHM* are similar. Consequently, it follows that

$$\frac{BG}{BH} = \frac{BH}{BM},$$

but since *BH* = *BE*, we get the following proportion:

$$\frac{BG}{BE} = \frac{BE}{BM}.$$

We can then establish a similarity between triangles *GEB* and *EMB*, since they both share a common angle *MBE* and the sides including this angle are in proportion. Since we can show that triangle *GEB* is isosceles, we also then know that triangle *EMB* must also be isosceles. Therefore *EM* = *MB*. Line *CM* is thus the perpendicular bisector of line segment *EB*. We may, therefore, conclude that point *M* must lie on line *CD*. Thus, we have demonstrated that the point *M* is at the intersection of the lines *AB* and *CD*.

Although this previous discussion was rather complicated, it used nothing more than elementary geometry, and, as a result, showed that the five possible constructions that can be created with an unmarked straightedge and a pair of compasses can also be made with just a pair of compasses alone. As we mentioned earlier, these Mascheroni constructions can also be attributed to the Danish mathematician Mohr—representing a misattribution. This does, on occasion, occur in mathematics, especially when, as is typical in the Western world, we look at the history of mathematics through European eyes. Another example of this is the famous Pythagorean theorem. We attribute this finding to the Greek philosopher Pythagoras, who flourished a few centuries after other citings of this famous relationship, such as the *Sulva Sutra*, which was written by the Indian mathematician Baudhayana in about 800 BCE, where there is a reference made to this geometric relationship—yet we still call it the *Pythagorean* theorem. Let's bear in mind, though, that just as Pythagoras likely did not know about Baudhayana's earlier discovery, Mascheroni wrote *Geometria del compasso* without any knowledge of the previous work by Mohr; therefore, we still give them credit for this marvelous finding.

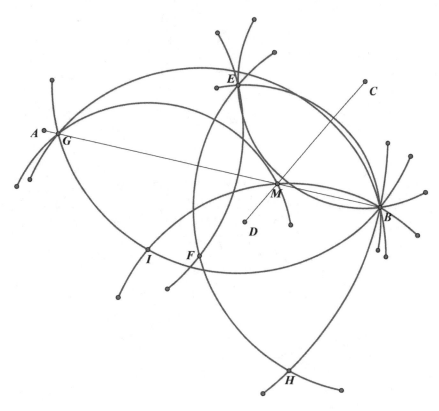

Figure 26.14

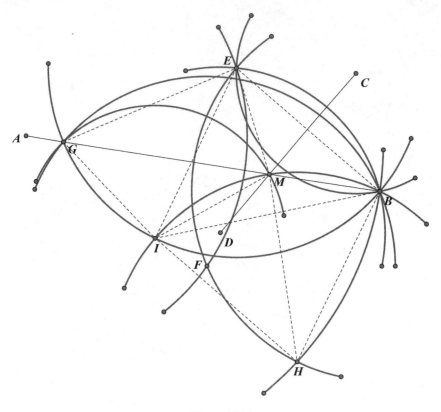

Figure 26.15

CHAPTER 27

Joseph-Louis Lagrange:
French/Italian (1736–1813)

It is not very well known that the famous French/Italian mathematician Joseph-Louis Lagrange was already fifty-one years old when he moved to France. Born as Giuseppe Luigi Lagrangia in Turin, Italy, he spent the first three decades of his life in Italy and the next twenty years in Germany, as Leonhard Euler's successor at the Berlin Academy of Sciences. However, in 1787, he accepted an offer of Louis XVI to move to Paris and become a member of the French Academy of Sciences. He remained in Paris for the rest of his life and is mostly remembered as Joseph-Louis Lagrange, although in Italy he is known as Giuseppe Luigi Lagrange, changing his Italian first and middle names to the French version to match with his French last name (see fig. 27.1).

Yet his important role in the introduction of the metric system during the French Revolution, together with the fact that his magnum opus, *Mécanique analytique*, was first published in Paris (although written in Berlin), may also contribute to his being perceived and remembered as French—except in Italy, of course. Although it would be perfectly right to call him an Italian mathematician, his family has a branch in France as well. Lagrange's paternal great-grandfather was a French army officer who had moved to Turin, which at that time was the capital of the Duchy of Savoy, to work for the Duke of Savoy. He married an Italian and the family stayed in Turin.

In 1720, Turin became the capital of the kingdom of Sardinia, and on January 25, 1736, Lagrange was born as the first of eleven children of

Figure 27.1. Memorial tablet in Turin.

Giuseppe Francesco Lodovico Lagrangia and his wife, Teresa. Only two of their children survived to adulthood. Lagrange's father was treasurer of the Office of Public Works and Fortifications in Turin.[1] Although his position would have been paid well enough to allow his family some degree of wealth, he unfortunately lost most of his money and most of his property with financial speculation. He wanted his eldest son to become a lawyer, and Lagrange accepted this wish without any hesitation. He studied at the University of Turin, where classical Latin would become his favorite subject. Initially he didn't show any interest in mathematics and found the subject rather boring. However, his mind suddenly changed when he accidentally came across a paper on the use of algebra in optics, written by the English astronomer and mathematician Edmond Halley (1656–1742). Halley is famous for computing the periodicity of a comet he had observed in 1682. The comet was named after him upon its predicted return in 1758. With his interest in mathematics aroused by Halley's memoir, Lagrange wanted to learn more and began to read mathematical texts on his own, including works by Maria Gaetana Agnesi and Leonhard Euler. His enthusiasm grew, and he threw himself into mathematics. After only one year of intense studying, he was an essentially self-taught but accomplished mathematician. In 1754, Lagrange published his first mathematical work, an analogy between the binomial theorem and the successive derivatives of the product of functions. This work, written in the form of a letter to the mathematician Giulio Fagnano (1682–1766), was not a masterpiece and showed that Lagrange was working alone, without the advice of a mathematical supervisor. Shortly after the publication of his paper, Lagrange discovered that the results were already contained in a published correspondence between Johann Bernoulli and Gottfried Wilhelm Leibniz. Lagrange was shocked by this discovery and feared that he would be accused of plagiarism. However, the bumpy start of his career as a mathematician actually increased his motivation, as he then wanted to prove as soon as possible that he was able to achieve his own significant results in mathematics. He began working on the tautochrone curve, "the curve for which the time taken by an object sliding without friction in uniform gravity to its lowest point is independent of its starting point."[2] The tautochrone problem had been solved by Christiaan Huygens 1659. Using geometry, he identified the curve to be an inverted cycloid, the curve traced by a point on the rim of a circular wheel as the wheel rolls along a straight line without slipping. Lagrange was able to provide a purely analytic solution to the tautochrone problem, which made his name known in the mathematical community. Moreover, he found a general method to find curves minimizing or maximizing certain quantities depending on the whole curve. For example, for each curve connecting a point A with a lower point B, one can calculate the time it would take an object sliding down from A to B along this curve in uniform gravity. Considering this time as a function of the curve, Lagrange's analytic method allows one to derive a set of equations that must be satisfied by the "shortest time" curve between A and B. Beyond the fact that Lagrange was able to solve the tautochrone problem without any geometrical arguments, his method also provided a much more general framework to formulate and investigate similar problems. He sent his results to Euler, who was very impressed. Euler had been working on similar problems, using related ideas, but Lagrange's approach considerably simplified and generalized Euler's earlier analysis. This general framework is now called *calculus of variations*, and the equations defining the maximizing or minimizing curve are called the *Euler-Lagrange equations*. In 1755, Lagrange, who was still a teenager, was appointed professor of mathematics at the Royal

Artillery School in Turin. Euler, who recognized Lagrange's mathematical talent and originality, convinced the president of the Berlin Academy of Science to create a position for Lagrange in Berlin, to lure him away from Turin. However, Lagrange wanted to stay in Turin, even though the position in Berlin would have been much more prestigious. He politely refused the offer but was happy to be elected to a corresponding member of the Berlin Academy in 1756. Two years later, Lagrange was one of the founders of a scientific society in Turin, which would become the Royal Academy of Sciences of Turin. In the following years, he published most of his writings in the transactions of the Turin Academy, known as the *Miscellanea Taurinensia*. His works were influenced by and extended those of Isaac Newton, Daniel Bernoulli, and Euler. He applied his new mathematical methods to many problems in physics, such as the propagation of sound, fluid mechanics, and the calculation of the orbits of Jupiter and Saturn. In his Turin papers, Lagrange made seminal contributions to the theory of vibrating strings and introduced what is now known as the *Lagrangian function* of a physical system; the Lagrangian function, often referred to as "the Lagrangian," is a central object in theoretical physics. In 1764, Lagrange was awarded the prize of the French Academy of Sciences for his work on the libration of the moon, which is a perceived oscillating motion of the face the moon presents to the Earth. On year later, another attempt, by Jean le Rond d'Alembert, to persuade Lagrange to leave Turin for a better position in Berlin failed. Lagrange's response to the offer indicates that he did not want go to Berlin, because there he would always be the second-best mathematician after Euler, a role to which he did not aspire. However, in 1766, an exorbitant offer by Catherine the Great persuaded Euler to return to St. Petersburg, Russia, where he had already held a position from 1727 to 1741. With Euler out of Europe, Frederick II himself, king of Prussia, wrote to Lagrange that he wanted to have "the greatest mathematician in Europe" at his court. Finally, Lagrange accepted the invitation and succeeded Euler as director of Mathematics at the Berlin Academy of Science. Frederick II had some disagreements with Euler that may have facilitated Euler's decision to go to St. Petersburg. After Lagrange's appointment as the new director of Mathematics, he disrespectfully wrote to the French mathematician Jean le Rond d'Alembert that he had "replaced a one-eyed geometer by a two-eyed one." In 1767, Lagrange married his cousin Vittoria Conti. He stayed in Berlin for twenty years, and Frederick II was indeed very pleased with his court mathematician. Lagrange produced a continuous flow of excellent papers covering the stability of the solar system, mechanics, dynamics, fluid mechanics, and probability. In a long series of papers extending over more than a decade, he basically created the theory of partial differential equations. The prize from the Paris Academy of Sciences was awarded to Lagrange on an almost-regular basis: He won the prize in 1766, for work on the libration of the moon; he shared the 1772 prize with Euler, for their work on the three-body problem;[3] he won the prize in 1774, again for his work on the motion of the moon; and he won the 1780 prize, this time for his work on the planets' perturbations of the orbits of comets.[4] While applications in classical mechanics and astronomy still played a major role in his research at the Berlin Academy, he also worked on number theory, proving in 1770 that every positive integer is the sum of four squares. The years in Berlin were the most productive in Lagrange's life: He was exempt from teaching and could devote all of his time to mathematics. It took some time until Italy realized Lagrange's mathematical genius and acknowledged that his leaving Turin for Berlin was a tremendous loss for his hometown. Upon Lagrange's visit in Paris in 1763, d'Alembert wrote ". . . in him Turin possesses a treasure whose worth it perhaps does not know." Occasionally, efforts were made to get Lagrange back to Italy, but Lagrange turned down generous offers; he sought neither wealth nor power, and wanted only to have peace to do mathematics, without any other obligations. Around 1780, Lagrange started to write his magnum opus, the previously mentioned *Mécanique analytique*: a single, comprehensive treatise containing his and his contemporaries' contributions to mechanics. Newton's presentation of mechanics in his famous *Philosophiæ Naturalis Principia Mathematica* (commonly known as the *Principia*) was based on geometrical methods, and Lagrange's intent was to transfer Newtonian mechanics and the art of solving problems in mechanics from a predominantly geometrical reasoning to a purely analytic and algebraic approach. His method was based on a set of general equations from which all the equations necessary for the solution of a particular problem can be derived. He wrote:

> No diagrams will be found in this work. The methods that I explain require neither geometrical, nor mechanical, constructions or reasoning, but only algebraic operations in accordance with regular and uniform procedure. Those who love Analysis will see with pleasure that Mechanics has become a branch of it, and will be grateful to me for having thus extended its domain.

Lagrange not only summarized all of the work done in the field of mechanics since the time of Newton but also dramatically simplified the application of Newton's theory via the use of differential equations, essentially condensing Newtonian mechanics into a single formula and eliminating any necessity for geometrical reasoning. However, Lagrange's monumental work *Mécanique analytique* was not published until 1788. At that time, he had already left Germany. In 1783, after years of illness, his wife, Vittoria, died; Lagrange became very depressed. Three years later, he also lost his patron, Frederick II. As a result of these losses, Berlin had become a less welcoming place for him, and he lacked any reason to stay. Many states in Europe saw their chance to hire him; the best offer came from France and included a clause that exempted Lagrange from any teaching obligations. In 1787, at age fifty-one, Lagrange left Berlin and was appointed to a paid position at the French Academy of Sciences in Paris, where his *Mécanique analytique* was published in two volumes, in 1788 and 1789. However, neither the new environment nor the publication of his great work could cheer up Lagrange; he was still very melancholic, and the printed copy of his *Mécanique* lay on his desk, unopened, for more than two years.

Figure 27.2. Joseph-Louis Lagrange.

When Lagrange came to Paris, the French Revolution was just about to start. In 1790, Lagrange was made a member of the committee of the Academy of Sciences, to standardize weights and measures. The existing system of measures had become impractical for trade and needed to be replaced. As the revolution developed, politics changed rapidly and the situation of anyone considered part of the establishment became potentially dangerous. In dealing with these circumstances, Lagrange gradually overcame his depression. Although he had already prepared his escape from France, it turned out that he would never face real danger. All foreigners born in enemy countries, including members of the Academy of Sciences, were subject to arrest once the Reign of Terror began in 1793. Fortunately, the famed chemist Antoine Lavoisier (1743–1794) intervened on behalf of Lagrange, and he was granted an exception. In the political turmoil, no one could feel safe, as one could be declared an enemy of the regime overnight. The weights and measures commission was allowed to continue, but soon several prominent figures—such as Lavoisier himself, mathematician Pierre-Simon Laplace, and physicist Charles-Augustin de Coulomb (1736–1806)—were thrown off the commission, while Lagrange became its chairperson.[5] In a trial that lasted less than a day, a revolutionary tribunal condemned to death Lavoisier and twenty-seven others. On the death of Lavoisier, who was guillotined on the afternoon of the day of his trial, Lagrange said:

> It took only a moment to cause this head to fall and a hundred years will not suffice to produce its like.

After Lavoisier's death, it was largely due to Lagrange's influence that the final choice of the unit system of meter and kilogram was settled and the decimal subdivision was finally accepted by the commission. Lagrange spent his last years as professor at the École Polytechnique. In 1808, Napoleon Bonaparte named Lagrange to the Legion of Honor and Count of the Empire. In April 1813, Lagrange died; that same year, he was buried in the Panthéon in Paris. He is also "one of the seventy-two prominent French scientists who were commemorated on plaques at the first stage of the Eiffel Tower when it first opened" in 1889.[6] Because many concepts and methods in both mathematics and physics bear his name, Lagrange is still well known. Among these are the Lagrangian function, Lagrangian mechanics, and the Euler-Lagrange equations, just to name a few.

CHAPTER 28

~

Sophie Germain:
French (1776–1831)

Perhaps the most significant mathematical achievement of the twentieth century was announced in June 23, 1993, by Andrew Wiles, at a lecture in Cambridge; there, Wiles claimed to have proven Fermat's Last Theorem, the 350-year-old claim (or conjecture) by the famous French mathematician Pierre de Fermat (see chap. 15). This made front-page news in the *New York Times* the next day, under the heading, "At Last, Shout of 'Eureka!' in Age-Old Math Mystery."[1] However, soon thereafter a slight error was detected in the proof. Over the next year, Wiles set out to correct this loophole, announcing its correction on September 19, 1994. It must be noted, though, that the technique and subject matter that Wiles used to prove Fermat's Last Theorem was known neither during Fermat's time nor during the ensuing centuries following his profound statement.

For centuries, mathematicians have grappled with this theorem. In the early nineteenth century, Sophie Germain produced what is believed to be one of the most significant contributions to dealing with Fermat's Last Theorem. Before we delve into Germain's significant contribution toward the solution of this perplexing problem, we need to understand the complex lifestyle she had to endure in order to pursue her love of mathematics.

Marie-Sophie Germain was born into a rather wealthy family on April 1, 1776, in Paris, France.[2] Her father was a successful silk merchant and politician who was able to support Sophie financially throughout her

Figure 28.1. Sophie Germain, 1790.
(Illustration from *Histoire du socialisme*, ca. 1880.)

entire adult life. Because one of her sisters and her mother shared the name Marie as the first part of their first names, she dropped it and became known as Sophie Germain. Germain was forced to stay home as a result of the unrest and street riots after the fall of the Bastille in 1789. Although we cannot be sure, it is commonly understood that she first encountered mathematics while reading some books in her father's library, particularly those on the history of the subject. From there, Germain taught herself both Latin and Greek so that she could read the works of Isaac Newton and Leonhard Euler. In this effort, she faced opposition from her family members, who held that the study of mathematics was inappropriate for women. Despite that familial opposition, she secretly continued pursuing her genuine interest in mathematics. Eventually, her mother became sympathetic and chose to support Germain's enthusiasm for mathematics.

In 1794, the École Polytechnique was established, but it would not admit women; however, lecture notes were available to anyone who requested them. Sophie, then aged eighteen years, acquired these notes and read them intensively. She then sent her observations of them (under a pseudonym, M. LeBlanc) to the famous Italian mathematician Joseph-Louis Lagrange, who was then a member of the faculty. Lagrange was very impressed with her submissions and requested a meeting with her; at that point, she revealed that she was a woman. This did not disturb Lagrange, and he continued to mentor her at her home.

Germain's interest specifically in number theory began in 1798, when she read the French mathematician Adrien-Marie Legendre *Essai sur la théorie des nombres*. Germain began corresponding with Legendre providing some brilliant ideas, which, ultimately, led him to include some of her work in his subsequent publication, *Théorie des Nombres*, with a citation to her for the ingenious aspect of her contribution.

Her interest in number theory was further motivated when she read the German mathematician Carl Friedrich Gauss's monumental work *Disquisitiones Arithmeticae*; once again using her earlier pseudonym of M. LeBlanc, she wrote to Gauss on November 21, 1804. In this correspondence, she presented some ideas on solving Fermat's Last Theorem. Gauss responded to her but did not comment on her work.

At this point, it would be helpful to recall the definition of Fermat's Last Theorem, which Fermat wrote in the margin of one of his arithmetic books in 1637. Writing that the margin space was insufficiently large for him to produce the proof of this statement, Fermat claimed that for integers $n > 2$, the equation $a^n + b^n = c^n$ cannot be solved with positive integers a, b, and c.

Sophie Germain made a few discoveries in the process of trying to prove Fermat's Last Theorem. Among them was that if $a^5 + b^5 = c^5$, then at least one of the variables, a, b, or c, must be divisible by 5. Furthermore, she took a special case by letting n be any odd prime number in $a^n + b^n + c^n$. She then claimed that if there exists another prime number $P = 2kn + 1$, where k is any positive integer not divisible by 3 such that $a^n + b^n - c^n = 0(\mathrm{mod}\ P)$, then P divides abc, and n is not an nth power residue (mod P). Finally, she concluded that Fermat's Last Theorem holds true for all values of n that do not divide a, b, or c. This is known as Sophie Germain's theorem, which was a major step—especially at the time of its development—toward proving Fermat's Last Theorem. She went ahead to show that her idea held for all odd primes $n < 100$. Later investigations into her work showed that it actually could be taken further, for every exponent $n < 197$. Germain continued to pursue a proof of this theorem in unpublished works, further motivating famous mathematicians such as Legendre and Lagrange.

Germain's name stays prominent in number theory not only for her theorem but also because there are some numbers named after her. One example is the Sophie Germain prime numbers; a prime number p is considered a Sophie Germain prime only if $2p + 1$ is also a prime number. For example, the number 3 is a Sophie Germain prime, since 3 is a prime number and $2 \cdot 3 + 1 = 7$, which is also a prime. On the other hand, the number 7 is not a Sophie Germain prime; this is because $2 \cdot 7 + 1 = 15$, which is not a prime. Her name is also found in algebra, where we have a Sophie Germain identity. This holds that for any values of x and y,

$$x^4 + 4y^4 = \left(\left(x+y\right)^2 + y^2\right)\left(\left(x-y\right)^2 + y^2\right) = \left(x^2 + 2xy + 2y^2\right)\left(x^2 - 2xy + 2y^2\right)$$

Germain continued to correspond with the most famous mathematicians of her time, one of whom, as we mentioned earlier, was Carl Friedrich Gauss. Gauss was one year younger than Germain, and because he had been corresponding with her under her pseudonym, he had assumed she was a man. During the Napoleonic wars, when Germain heard that the French were occupying Gauss's hometown of Braunschweig, Germany, she wrote to a French army general who was a family friend, to ensure Gauss's safety.[3] When Gauss found out that his protection was sought by Sophie Germain, he was astonished to learn that all the while he was corresponding with M. LeBlanc, he had actually been in contact with a woman. Gauss then went forward to praise Germain's genius heartily. Interestingly, Gauss and Germain never met in person.

Although Germain was most interested in mathematics and number theory, her concentration was not limited to only this field of study. She also wrote an award-winning paper on elasticity.[4] Germain published her prize-winning essay, "Récherches sur la théorie des surfaces élastiques," at her own expense in 1821 (see fig. 28.2). She did this mainly because she wanted to present her work in opposition to that of the French mathematician Siméon-Denis Poisson. Beyond mathematics, number theory, and elasticity, Sophie Germain also studied philosophy, psychology, and sociology. Much of her work was published posthumously, including her "Mémoire sur la courbure des surfaces," a piece she wrote on elasticity, using the mean curvature in her research.

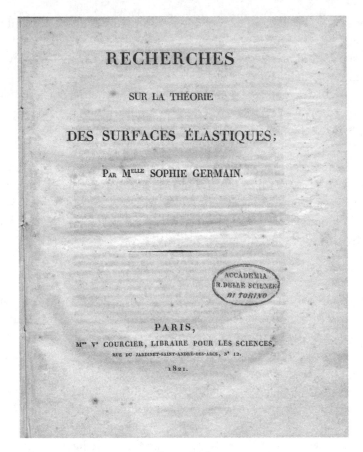

Figure 28.2.

In 1829, Germain began to suffer from breast cancer; despite the pain, she continued to pursue her work feverishly, publishing papers until her death, on June 27, 1831.[5] Germain died in the house where she lived her entire life—the house that today bears her name—at 13 rue de Savoie. She was further honored posthumously, by Gauss when he indicated that the University of Göttingen should have bestowed an honorary degree upon her. However, this was six years after her death. Essentially, Germain's life story is that of a woman of means who did not allow her gender to stop her from pursuing mathematics and scientific research, despite the opposition from her family and culture on account of her sex.

Figure 28.3.

Although she was not widely recognized for her brilliance during her own lifetime, Sophie Germain has had many honors bestowed upon her since her death in 1831. A street in Paris, Rue Sophie Germain, carries her name (see fig. 28.3), and a statue of her stands in the courtyard of the Paris school that also bears her name: École Sophie Germain. As was mentioned above, the house at 13 rue de Savoie is named for her, and it has been designated as a historical landmark. Furthermore, the Sophie Germain Hotel is located at 12 Rue Sophie Germain. On account of not only these physical landmarks and locations but also the Prix Sophie Germain, a mathematics prize that is offered annually by the Paris Academy of Sciences, this female polymath remains in the public eye today.

~

CHAPTER 29

Carl Friedrich Gauss:
German (1777–1855)

It is not easy to summarize the seventy-eight-year-long life of one of the greatest mathematicians of all time, but we hope to provide an overview of this great man. Carl Friedrich Gauss was born on April 30, 1777, in Braunschweig, Germany, in a poor, working-class family (see fig. 29.1). There are many stories of how it was determined early on that Gauss was a child prodigy. It has been said that at the age of three, he was able to find a mistake in his father's household calculations.

Perhaps the most famous story of Gauss's prodigious youth is that when he was eight years old, his elementary-school teacher, in an effort to keep the class busy while he did some clerical work, asked the class to add the numbers from 1 to 100. No sooner had he given the assignment than young Gauss put his slate down, indicating that he had arrived at this requested sum. The teacher ignored him so as to let the rest of the class complete the assignment. After a half hour, when the teacher sought the results from his students, Gauss was the only one who had the correct answer. His clever method for finding the sum so quickly was to add the numbers—but not as the rest of class did (1 + 2 + 3 + 4 + 5 + . . . + 98 + 99 + 100). Rather, he added the first and the last numbers (1 + 100 = 101), then he added the second and the next-to-last (2 + 99 = 101), and then he continued along the same pattern (3 + 98 = 101, etc.) until all numbers in the list were coupled. He realized that there were fifty pairs of such additions, and so all he needed to do to get the sum was to multiply 50 by the repeated sum of 101, that is,

Figure 29.1. Carl Friedrich Gauss.
(Oil on canvas, by Christian Albrecht Jensen, 1840.)

$50 \cdot 101 = 5{,}050$. This is a story that all good math teachers should be sharing with their students at the appropriate time, that is, when introducing the formula for the sum of an arithmetic sequence.

In 1791, when Gauss was fourteen years old, Carl Wilhelm Ferdinand, the duke of Braunschweig, discovered Gauss's brilliance and offered to finance his study at what is known today as the Braunschweig University of Technology. Upon graduation there, Gauss was admitted to the University of Göttingen, still financially supported by the duke. He studied at Göttingen from 1795 until 1798, whereupon he left without a degree. However, during this period of time, he wrote and published his monumental work, largely on the theory of numbers, *Disquisitiones Arithmeticae*. Although it was completed in 1798, due to publishing difficulties in Leipzig, it was not published until 1801.

Disquisitiones Arithmeticae is often considered his greatest masterpiece. Even though arithmetic was Gauss's favorite subject, he also delved into astronomy, geodesy, and electromagnetism—among other fields as well. The book has seven sections; the first three deal with the theory of "congruences," or as we tend to call it today modular arithmetic. The fourth section deals with the theory of quadratic residues, for which he had an ingenious approach that amazed and impressed many people in his time. The study of quadratic equations continued until the seventh section, which most people consider the highlight of the book. In this section, he discussed the equation $x^n = 1$, where n is a given integer. In his discussion, he combined arithmetic, algebra, and geometry. This equation is the basis for the algebraic approach to the problem of constructing a regular polygon of n sides. This is one of Gauss's proudest discoveries, namely that a regular polygon of seventeen sides can be constructed using only an unmarked straightedge and a pair of compasses. This was one of the major advances in geometry since the time of famous Greek mathematicians. Gauss often said that he would like to see this seventeen-sided polygon on his gravestone, but the stonemason balked at the suggestion, saying that drawing such a figure would look like a circle with seventeen points on it.

Today we know that a regular polygon of n sides is constructible with an unmarked straightedge and a pair of compasses if n is equal to any of the following: 3, 4, 5, 6, 8, 10, 12, 15, 16, 17, 20, 24, 30, 32, 34, 40, 48, 51, 60, 64, 68, 80, 85, 96, 102, 120, 128, 136, 160, 170, 192, 204, 240, 255, 256, 257, 272, 320, 340, 384, 408, 480, 510, 512, 514, 544, 640, 680, 768, 771, 816, 960, 1020, 1024, 1028, 1088, 1280, 1285, 1360, 1536, 1542, 1632, 1920, 2040, 2048, . . .

We know this because a regular n-gon is constructible with these tools if and only if $n = 2^k p_1 p_2 \ldots p_t$, where k and t are non-negative integers, and the p_i's (when $t > 0$) are distinct Fermat primes, which are prime numbers that can be expressed as $2^{2^n} + 1$. The five known Fermat primes are $F_0 = 3$, $F_1 = 5$, $F_2 = 17$, $F_3 = 257$, and $F_4 = 65537$.

After returning to the University of Göttingen, Gauss finally received his first degree in 1799. The duke further requested that Gauss submit a doctoral dissertation to Helmstedt University, whereupon he then received his doctorate. In the years following, he delved in astronomy to help establish an observatory at Göttingen. On October 9, 1805, Gauss married Johanna Osthoff, with whom he had a son and a daughter. His wife died on

October 11, 1809, and their second child died shortly thereafter. The following year, Gauss married Minna Waldeck, with whom he had three more children. During this marriage he grew very close to his children. Sadly, his second wife died in 1831.

Returning to Gauss' academic career, in 1807 he left Brunswick and arrived in Göttingen to assume the position of director of the observatory. The following year there began a series of misfortunes for Gauss. His father died in that year, and this coupled with the death of his wife two years later, caused him to become quite depressed. His production continued despite these unfortunate events. In 1809, he published a two-volume treatise on the motion of celestial bodies. The first of the two volumes covered differential equations, conic sections, and elliptical orbits; the second concentrated on estimating a planet's orbits.[1]

In 1818, he accepted the task of developing a geodesic survey of the state of Hanover so that it could then be linked with the Danish grid. Here, once again, his incredible ability to do calculations mentally was a great help. Many of his discoveries seem to have resulted from his ability to do mental calculations far beyond those that an average person can conceive.

One of his discoveries published in *Disquisitiones Arithmeticae* is an example of this. What is now referred to as Gauss's Eureka theorem (because he wrote in his diary, "EΥPHKA! num = $\Delta + \Delta + \Delta$")[2] is that every positive integer can be expressed as the sum of triangular numbers. Triangular numbers are 0, 1, 3, 6, 10, 15, . . ., and they can be expressed as $\dfrac{n(n+1)}{2}$. For example, $18 = 15 + 3$, and $28 = 15 + 10 + 3$.

Another of Gauss's discoveries was that he proved what is today referred to as the fundamental theorem of algebra. In simple terms, it states that every algebraic equation in one variable has a root, or an answer. These roots can either be real or complex, and so Gauss used the notation of $a + bi$, where $i = \sqrt{-1}$. Furthermore, Gauss was the first to give a comprehensive explanation of complex numbers and their labeling as points on the plane with Cartesian coordinates.

For these reasons among many others, Gauss was considered one of the most brilliant mathematicians of his time. In 1816, the Paris Academy offered a prize for anyone who could prove Fermat's Last Theorem in the period 1816–1818. Gauss was urged to compete, but he wrote to a friend that "Fermat's Last Theorem as an isolated proposition has very little interest for me, because I could easily lay down a multitude of such propositions, one

could neither prove nor dispose of."[3] As you might recall, Fermat's Last Theorem states that no three positive integers a, b, and c can satisfy the equation $a^n + b^n = c^n$ for any value of n greater than 2. It took 358 years until
the British mathematician Andrew Wiles published the proof in 1995.)

It was well known that Gauss did not enjoy teaching; however, on occasion, he did announce a lecture or teach private lessons. For instance, see
the announcement from 1831 in figure 29.2, where he stated, "at 10 o'clock
I will explain the use of probability calculus in applied mathematics, especially astronomy, advanced geodesy and crystallometry. I will teach practical astronomy in most private sessions. The first lecture will be on October
28th." Latin was a favorite language for him to use for mathematics and
other scientific communication, as evidenced by this announcement.

Gauss endured some additional depressing times between 1817 and
1832, when his mother, who was always very dear to him, became ill and
lived with him until she died in 1839.[4]

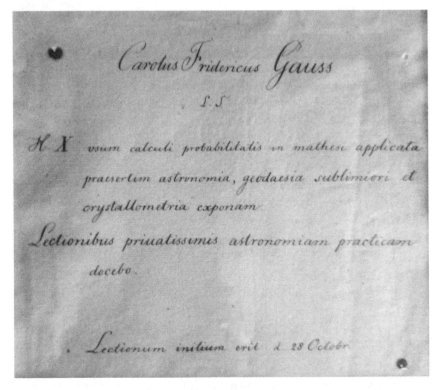

Figure 29.2. An 1831 announcement written by Gauss regarding a lecture and
private sessions he would be holding in October of that year.

In 1832, Gauss formed a partnership with a leading physicist of his time, Wilhelm Weber (1804–1891). They constructed the first electromagnetic telegraph in 1833. The first connection was between Gauss's magnetic observatory and the Institute for Physics in Göttingen, Germany. At the prompting of Prussian scientist Alexander von Humboldt, Gauss and Weber determined measurements of Earth's magnetic field in many regions of the world. At Gauss's magnetic observatory, they began to modify Humboldt's procedures, which did not please him. Yet Gauss's changes were far more effective and accurate. Throughout his life, Gauss was also involved with scientists in the life sciences, such as the famous German physician and anthropologist Johann Friedrich Blumenbach (1752–1840). Their connection is evidenced by a memo Gauss wrote to Blumenbach, shown in figure 29.3. In that memo, written in his hand on October 7, 1821 Gauss states, "*Beigehende interessante Abhandlung verehrtester Herr College ist mir von ihrem Verfasser aus München mit dem Ersuchen übersandt worden, solche der Königl. Societät als Zeichen seiner Verehrung zu überreichen.*" This can be translated as, "The enclosed interesting treatise, honored Colleague, has been sent to me by its author from Munich with the request that the Royal Society accept it as a token of his reverence."

In 1837, for political reasons, Weber had to leave Göttingen, after which Gauss's work gradually became less voluminous, yet he was always eager to support other scientists in their work. Carl Friedrich Gauss died on November 23, 1855, in Göttingen, and he is buried in the Albani Cemetery there.

Perhaps his most famous quote, and one that is often mentioned today is that "Mathematics is the queen of sciences, and arithmetic the queen of mathematics."

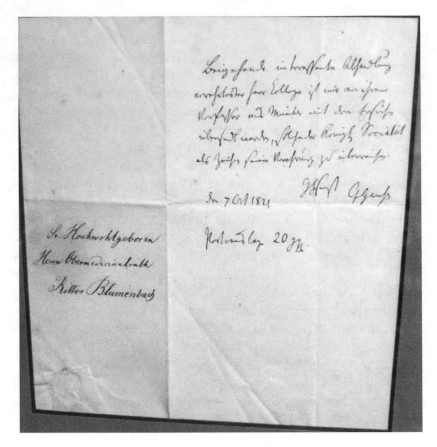

Figure 29.3. A memo from Gauss to
Johann Friedrich Blumenbach, dated October 7, 1821.

Figure 29.4. Gauss on this deathbed, 1855. (Painting by Philipp Petri, 1855.)

CHAPTER 30

~

Charles Babbage: English (1791–1871)

In today's modern world, the calculator and the computer are often taken for granted. However, we should look back to determine from where the concept of a calculator—or computer, as it was originally known—emanated. The honor of having first developed a machine that does calculation belongs to the English mathematician Charles Babbage, who was born in London on December 26, 1791.

He was the son of Benjamin Babbage, a London banker. Charles Babbage was educated mostly at home, since he was frequently ill. Even in these early days, he developed a love for mathematics. In 1810, he was accepted to study mathematics at Trinity College of Cambridge University, where in a short time he found himself more advanced than his instructors in mathematics. This prompted him to join a group of students also disappointed with the level of instruction. This group, called the Analytical Society, was devoted to exploring more-advanced issues in mathematics. During this time, he and his colleagues were disturbed by the inaccuracy of the numbers of logarithm tables. He felt better about calculating such values himself, which was his initial motivation to develop a machine that could do that task accurately. In 1817, he received his master's degree from Cambridge University. Soon thereafter, he worked there as a lecturer of mathematics.

By 1816, he was elected a fellow of the Royal Society, and which eventually led him to participate in the founding of the Royal Astronomical Society in 1820. This was about the time when his interest began to take him in

Figure 30.1. Obituary portrait of Charles Babbage, published in the *Illustrated London News*, November 4, 1871. (Portrait derived from a photograph of Babbage taken at the Fourth International Statistical Congress, London, July 1860.)

the direction of developing a calculating machine. This initial interest had been lurking in his mind for some time. In 1821, Babbage developed the Difference Engine No.1 (see fig. 30.2). In 1822 he announced his findings to the Royal Astronomical Society in England at a lecture he had given to a select group. Although Babbage conceived of having his machine print out the results, initially an assistant had to serve as a scribe to copy the numbers that were generated. The following year he received the gold medal from the Astronomical Society as a reward for his developing this amazing machine. The work on further improvement of the machine was generously funded by the government. Babbage did not get along well with the funders, since many of their questions seem to have been, by his measure, ridiculous. Eventually, funding was withdrawn because it took so long to reach a workable model. However, it must be said that during this time in 1827 Babbage had a most unfortunate year. His father died, his wife died and two of his children died. He was advised to take time off and spent the better part of a year traveling on the continent of Europe. He returned to his work in 1828. Despite the lack of financial support, his model was completed in 1832, and it was able to assist in compiling mathematical tables. Yet he was displeased with the limitations of this machine, and so he began to develop one that could do a wider variety of calculations. Babbage played a significant role in the establishment of the British Association for the Advancement of Science, as popularized through a rather controversial paper he wrote in 1830. As influential as Babbage was, work was stopped on the difference machine in 1834 because the government was displeased with the progress shown to date.

During his constructive years Babbage remained at Cambridge University, where he held the Lucasian Chair of Mathematics from 1828 to 1839 but never presented a single lecture. Babbage was a very active member of the intellectual society and supported research in many scientific areas, such as cryptography which was then used by the British and American governments. During his efforts to continue to improve the development of these calculating machines, in 1843, a Swedish inventor named George Scheutz (1785–1873) was able to construct the Difference Machine based on Babbage's design. In 1837, Babbage described a successor to the second version of the Difference Engine, and called it the Analytical Engine (fig. 30.3). This was a general-purpose computer whose design regarding memory was the forerunner of the electronic computers that followed, years later. The conceived memory was to hold 1,000 numbers composed of 40 decimal digits each. The machine was intended to perform the four arithmetic operations and square-root extraction. The programming language used was analogous to the modern-day assembly languages. Punch cards were used, one for arithmetic operations, one for numerical constants, and one for transferring numbers from storage to the arithmetic unit. Unfortunately, the machine was never successfully completed to the level Babbage desired, and it ran only a few tasks, with some obvious errors.

It is sad to note that Babbage died a frustrated man, since his visions were never fully realized, which he blamed on the government's failure to provide the proper financial support. After Babbage's death on October 18, 1871, his work was continued by his son Henry Prevost Babbage. In fact, even he had to do much of his own funding to continue his father's work. In 1910, Henry Babbage constructed his version of the Analytical Engine, which was not programmable and had no storage (see fig. 30.4).

Babbage's legacy includes many diverse contributions, such as having compiled reliable actuarial tables and having assisted in establishing the

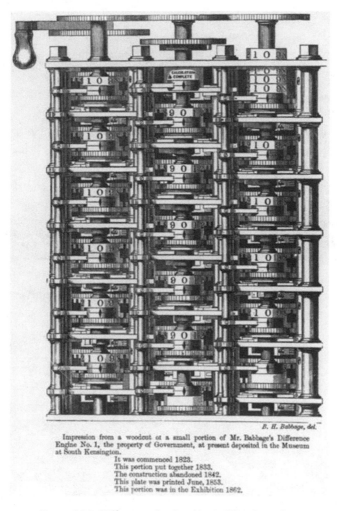

B. H. Babbage, del.

Impression from a woodcut of a small portion of Mr. Babbage's Difference
Engine No. 1, the property of Government, at present deposited in the Museum
at South Kensington.
It was commenced 1823.
This portion put together 1833.
The construction abandoned 1842.
This plate was printed June, 1853.
This portion was in the Exhibition 1862.

Figure 30.2. Difference Engine No. 1. (Woodcut after
a drawing by Benjamin Herschel Babbage, 1853.)

modern British postal system. He also invented a speedometer, occulting
lights for lighthouses which are lights that stay on longer than the period of
darkness, and a locomotive cow catcher, which is the metal device used by
trains to clear the tracks of obstacles that would interfere with rail traffic.
So in Charles Babbage we have the initiator of our computer world, who
struggled to make a machine whose invention was motivated by what he
saw as something desperately needed to correct earlier-accepted erroneous
information.

Figure 30.3. Analytical Engine. (Courtesy of Doron Swade.)

Figure 30.4. Henry Babbage's Analytical Engine Mill, built in 1910, on exhibit at the London Science Museum. (Wikimedia Creative Commons, author: Marcin Wichary, licensed under CC BY-SA 2.0.)

~

CHAPTER 31

Niels Henrik Abel:
Norwegian (1802–1829)

Norway is one of the wealthiest countries in the world. It has extensive natural resources and, on a per-capita basis, Norway is the world's largest producer of oil and natural gas outside the Middle East. In contrast to most other countries in the world, Norway has no foreign debt; the state revenues generated from the petroleum industry even allowed the government to establish a sovereign wealth fund, which, by 2017, has accumulated a value of $185,000 for each citizen.[1] Public healthcare and public education are virtually free, which are probably two reasons why Norway ranked first in the 2017 World Happiness Report of the United Nations.[2] While Norway is obviously a very good place to live in this day and age, and it attracts immigrants from all over the world, the situation was very different at the beginning of the nineteenth century. Frequent famines, induced by climatic extremes during the Little Ice Age, had led to great loss of life. For more than four hundred years, Norway had been trapped in an unequal union with Denmark and was essentially controlled by the Danish authorities in Copenhagen.

During these difficult times, Niels Henrik Abel was born in a small village on the West Norwegian coast on August 5, 1802. Niels was the second of seven children of Anne Marie Simonsen and Sören Georg Abel, a pastor. Political conflicts between Denmark and Britain, which had already precipitated the First Battle of Copenhagen in 1801, made the situation for the Norwegian population even worse in 1807, when Denmark entered into an alliance with Napoleon and the British fleet reacted by imposing a blockade

Figure 31.1. The only contemporary portrait of Niels Henrik Abel, painted by Johan Gørbitz in 1826.

on supply lines between Denmark and Norway. For several years, Norway was able to neither export nor import goods to and from the continent, which led to a severe economic crisis that culminated in mass starvation in 1812. Since Abel's parents could not afford to send their children to school, he was educated at home by his father, who had a degree in philology and theology. Records suggest that the difficulties of a childhood in a poor household were exacerbated by alcohol abuse of both parents.[3] However, fortunately, the general economic situation became slightly better when the Napoleonic Wars came to an end and Denmark lost its power over Norway. Norway took a chance and declared independence in 1814. At the age of thirteen, Abel entered the Cathedral School in Christiana (now Oslo). He soon wrote home that he "felt right in his element,"[4] yet he achieved only moderately satisfactory marks in his first year at school. In the nineteenth century, school corporal punishment was an accepted method of behavior management, and student injuries caused by physical punishment were not uncommon. Abel's mathematics teacher was dismissed in 1817 because he had beaten a student so hard that the student died eight days afterward.[5] Fortunately, the new mathematics teacher, Bernt Holmboë (1795–1850), who was only seven years older than Abel, recognized his student's great talent for and fascination with mathematics and began to mentor him. He encouraged him to study university-level books outside the school curriculum, and together they read the works of Leonhard Euler, Isaac Newton, Jean le Rond d'Alembert, Joseph-Louis Lagrange, and Pierre Simon Laplace. When Abel's father lost his job in 1818, he started drinking excessively, and he died two years later. His death dramatically increased the family's financial problems and put an additional burden on Abel, who was now responsible for his mother and his siblings, since his elder brother slipped into depression and could not support the family. Without the help of Holmboë, it would not have been possible for Abel to continue his education. Holmboë raised funds for Abel that allowed him to finish school and then enter the University of Christiania in the fall of 1821. In a report, Holmboë wrote about his gifted student:

> With the most incredible genius he unites ardor for and interest in mathematics such that he quite probably, if he lives, shall become one of the great mathematicians.[6]

The University of Christiania was founded in 1813, and initially it did not offer any studies in the natural sciences; its focus lay in the profession-oriented studies: theology, medicine, and law.[7] Already, as a freshman, Abel was one of the best mathematicians in the country. During his last year in school, he had begun pursuing his own mathematical ideas; in particular, he was interested in solving quintic equations, a major unresolved mathematics problem at that time. Quintic equations are equations of the form $ax^5 + bx^4 + cx^3 + dx^2 + ex + f = 0$, that is, polynomial equations in which 5 is the highest power of x occurring in the equation. Here, the coefficients a, b, c, d, e, and f represent real numbers, and x is the unknown quantity. You may recall that the roots of a quadratic equation $ax^2 + bx + c = 0$ can be obtained by the formula $x_{1,2} = \dfrac{-b \pm \sqrt{b^2 - 4ac}}{2a}$.

For cubic and quartic equations there exist similar, albeit much more complicated, formulas. The first explicit formula for the solution of a general quadratic equation is attributed to the Indian mathematician Brahmagupta

(598–ca. 668 CE). The formulas for cubic and quartic equations were developed by the Italian mathematicians Scipione del Ferro (1465–1526) and Lodovico de Ferrari (1522–1565) and were first published in a book by another Italian mathematician, Gerolamo Cardano (1501–1576). However, the problem of solving quintic equations of the most general form had occupied mathematicians for hundreds of years, and none of them had been successful. In 1821, Abel believed that he had solved the problem in full generality; he sent a paper to Ferdinand Degen (1766–1825) in Copenhagen, the leading mathematician of the northern countries at that time. Upon Degen's request to provide a numerical example of his method, Abel discovered a mistake. However, Degen noticed the brilliancy in Abel's mathematical reasoning and advised him to make use of his abilities in other areas of mathematics as well.

After one year at the university, Abel's grades were not very outstanding, except for mathematics, where he excelled. Since there were no advanced study programs other than theology, medicine, and law, Abel had to study mathematics entirely on his own by borrowing mathematics books from the library. He read all of the mathematical texts he could find. There were only two mathematics professors at the university, and they soon realized that Abel would have to go abroad for any further study. In 1823, they financed a trip to Copenhagen so that he could visit the mathematicians there. However, it turned out that he already knew everything they had shown him. At a ball in Copenhagen, Abel met Christine Kemp (1804–1862) who became his fiancée one year later and subsequently followed him to Norway, where she found a job as a governess. Meanwhile, Abel had published several papers on topics in advanced calculus in a new scientific journal founded by one of his professors. He took up his work on quintic equations again, using a different approach, and eventually he solved the centuries-old problem, yet in a very surprising way. He proved that there exists no algebraic solution of a general polynomial equation of degree 5 or higher; that is, he determined that it is impossible to express the solutions in terms of the coefficients of the equation—as was done with other higher-degree equations. Solutions do exist, but they can be calculated only by approximation methods; it is, in general, not possible to "solve for x," if the equation is of fifth or a higher degree. This important result is now known as the Abel-Ruffini theorem, since Paolo Ruffini (1765–1824) had published an incomplete proof in 1799. In order to do his proof, Abel developed (independent of Évariste Galois [see chap. 32]) the branch of mathematics known today as *group theory*. There is also a type of group named for Abel; the abelian group is one that aligns with the commutative property; in other words, the order of operations is irrelevant for the abelian group.

Abel applied for funding to travel to the centers of mathematics in France and Germany, but he received only a small stipend to learn the languages, with a promise that he would then receive a travel grant two years later. To have an impressive piece to his name in anticipation of his visit with the great mathematicians in Europe, he published at his own expense his work on equations of fifth-degree. He wrote the text in French to reach a larger audience, while at the same time shortening the proof as much as he could, to save printing costs. He sent the work to several mathematicians on the Continent, including Carl Friedrich Gauss (1777–1855), but the extremely condensed style of his writing made the proof very hard to read, and so his work did not receive the attention he had hoped for. Having gained a good knowledge of French and German, Abel felt well prepared for his trip to Europe, and he wrote a personal letter to the king of Norway to obtain the travel grant earlier. With a scholarship from the Norwegian government, he was able to start his journey to the Continent in the fall of 1825. Although the plan was to go to Göttingen to visit Gauss and then to visit the French Academy of Sciences in Paris (the world's center of mathematics at that time), Abel first went to Berlin. There, he met August Leopold Crelle (1780–1855), an engineer and mathematician with good contacts to the government and who had long been planning to found a German mathematics journal to challenge the dominance of the well-established French journals. Crelle encouraged Abel to write an expanded and more accessible version of his results on quintic equations, a masterpiece of mathematics he eagerly wanted to publish in his journal. The first issue of Crelle's *Journal für die reine und angewandte Mathematik* (Journal of Pure and Applied Mathematics) appeared in February 1826 and contained seven papers by Abel, who was also a principal contributor to the following issues. The high quality and importance of Abel's papers were essential to establish the journal's reputation. Today, Crelle's journal is still one of the most renowned journals of mathematics. Abel abandoned his plans to visit Gauss, when he was informed that Gauss didn't approve of his work. (Yet Gauss had actually never read Abel's work on the quintic equation, as it was discovered unopened after Gauss's death.)

When Abel came to Paris, he completed a large manuscript, which he considered to be his most impressive work so far, containing a whole new theorem on the addition of algebraic differentials with fundamentally new insights. He submitted it to the French Academy of Sciences for publication, and hoped that this publication in one of the most important journals of mathematics would make a strong-enough impression on the authorities in Norway to create a position for him at the university in Oslo. Augustin-Louis Cauchy (1789-1858) and Adrien-Marie Legendre (1752-1833) were appointed as referees. Abel spent the winter in Paris, awaiting an answer, but Cauchy had set the manuscript aside and forgotten about it. With almost no money left, eating only one meal per day, Abel's health deteriorated. He developed a fever and a cough but kept on working at an enormous pace, writing further papers for Crelle's journal. Although he frequented science circles in Paris and made the acquaintance of the leading mathematicians there, they had little interest in his work, and he never really had the opportunity to discuss his mathematical ideas with them. Many years later, the mathematician Joseph Liouville (1809-1882), who was a student when Abel visited Paris, said that meeting Abel without getting to know him was one of the greatest mistakes of his life. In spite of having published a number of papers in Crelle's journal, Abel's trip to Europe was a disappointment for him, and he became homesick.

Upon his return to Norway in 1827, he was rather depressed and poor. Without publishing in Paris, and having been unable to make contact with Gauss, he found that his grant was not renewed. He had to take a private loan to clear the debts of his family. He placed advertisements in newspapers as a private tutor, seeking to earn some money. Meanwhile, he continued to send papers to his friend Crelle at an incredible rate—most of them pioneering works in different fields of mathematics. Crelle tirelessly tried to use his influence to create a permanent position for Abel at the University of Berlin. In the spring of 1828, as Abel's financial situation improved, he obtained a temporary position as a lecturer at the university in Oslo, substituting for a professor who went on a scientific expedition to Siberia. Although his publications in Crelle's journal became increasingly more favorably recognized by the mathematicians in the French Academy, there was still no hope for a permanent position in Norway. His health condition had not really improved since he had left Paris, and it worsened in the fall of 1828. He wanted to spend Christmas with his fiancée, who worked as a governess in Froland, a district more than 250 kilometers from Oslo. Abel had to travel by sled in the bitter, cold Norway winter, which would have been an extremely exhausting trip even for a strong and healthy person.

When he arrived, he was seriously ill. At Christmas, he felt a little better and could enjoy the celebration, but soon he was bedridden again, becoming weaker and weaker. Fearing that his greatest work, the paper submitted to the French Academy, had been lost forever, Abel invested all of his remaining physical energy to write down the proof of the main theorem on the addition of algebraic differentials again. After a violent hemorrhage, Abel was diagnosed with tuberculosis, from which he had probably been suffering since his earlier stay in Paris. He died on April 6, 1829, at the age of twenty-six. Two days later, not aware of Abel's death, Crelle sent a joyful letter to Abel to tell him that a permanent position as a full professor was awaiting him at the University of Berlin. He wrote:

> As far as your future is concerned, you can now be completely at ease. You belong among us and are secure. . . . You will be coming to a good country, to a better climate, closer to science and to sincere friends who appreciate you and are fond of you.[8]

One year later, Abel was posthumously awarded the Grand Prix of the French Academy of Sciences for his outstanding achievements in mathematics. After intensive searching, Cauchy finally found Abel's monumental "Paris memoir," which was then first published in 1841; it still stands as a milestone in the development of mathematics. In his short and tragic life, Niels Henrik Abel made deep and influential mathematical discoveries. Several mathematical theorems, equations, and objects bear his name; even a crater on the moon was named after Abel. When the Norwegian mathematician Sophus Lie (1842–1899) learned that Alfred Nobel's plans for annual prizes would not include a prize in mathematics, he proposed creating an Abel Prize for outstanding achievements in mathematics, to be awarded annually, beginning in 1902. However, with Lie's death in 1899, the motor behind these plans was gone and, also for financial reasons, the government decided to erect an Abel monument instead of funding an Abel Prize. In the late 1960s, oil exploration in the North Sea started and Norway's Oil Age began, turning it into one of the wealthiest countries in the world by the end of the millennium. With money in abundance from the oil industry and with the 200th anniversary of Abel's birth approaching, the Norwegian government finally established the Abel Prize in 2001. The Abel Prize was awarded with prize money amounting to 6 million Norwegian kroner (approximately $750,000). Together with the Fields Medal, the Abel Prize is viewed as the highest honor a mathematician can receive.

CHAPTER 32

Évariste Galois:
French (1811–1832)

You might expect the life story of Évariste Galois to be a short one, since he lived only to age twenty. Yet during these two decades years, he experienced a number of turbulent events. Before we consider his biography, we should note that his main contribution to mathematics is an entire field of study that bears his name—something not particularly common in the field of mathematics. Galois theory is a part of abstract algebra and draws a connection between two major theories: group theory and field theory. To the nonmathematician, this explanation might seem meaningless. However, we will try to show some of the new insights that result from this theory. For example, Galois's work allows us to determine solutions to higher-degree equations using only the four arithmetic operations and extractions of radicals (such as square roots, cube roots, etc.). His work also allows us to determine which regular polygons are constructible using only a straightedge and compasses, as well as why it is not possible to trisect a general angle using only a straightedge and compasses. These are just a few of the rather simple topics that might well have been presented in your secondary-school curriculum and that owe their solution to Galois's work.

Let us now consider how this mathematical genius navigated his twenty years. He was born on October 25, 1811, in Bourg-le-Reine, France, where in 1814 his father, Nicolas-Gabriel Galois, became mayor of the town. Uncommonly for the times, his mother, Adélaïde-Marie Demante, was a highly educated lawyer and provided her son home schooling until age twelve,

Figure 32.1. Évariste Galois. (Drawing on gray paper, ca. 1826.)

even though at age ten, Évariste had been offered admission to the College of Reims. In 1823, he entered the Lycée Louis-le-Grand. There he won first prize in Latin, but much preferred studying mathematics, which he did intensively at age fourteen. He showed his talent by reading rapidly through Adrien-Marie Legendre's *Éléments de Géométrie*, a book that, in a certain sense, served as the model for the American high-school geometry course. At age fifteen, he started to take the theory of equations very seriously. Curiously, his teachers were not impressed with him, or, as some might say, they felt intimidated by him.

In 1828, he applied to the prestigious École Polytechnique but did not perform well enough on the oral exams to be accepted. Shortly thereafter, he applied for admission to the École Normale, an inferior institution, and was accepted; the examiners seemed to be impressed by him. In 1829, he published his first paper on the topic of continued fractions. These are fractions of the form

$$a+\cfrac{b}{c+\cfrac{d}{e+\cfrac{f}{g+\cdots}}}\ ;$$

and they can be used to express such numbers as

$$\sqrt{2}=1+\cfrac{1}{2+\cfrac{1}{2+\cfrac{1}{2+\cfrac{1}{2+\cdots}}}}\ .$$

Some of the papers he submitted shortly thereafter were not accepted, for a variety of reasons, some of which were political and not necessarily mathematical. Paris was rather turbulent in January 1831. Galois quit school to join a militia, where split his time engaged in politics and in mathematics. Occasionally, members of the militia group were arrested, but he had no long stay in jail. In April 1831, Galois and the other officers of the militia were acquitted of all charges; and later, in May, they were honored with a banquet. At this banquet, Galois proposed a toast to the king—actually threatening his life; consequently, Galois was arrested again the following day. In June of that same year, he was acquitted once again. His radical behavior continued right after Bastille Day (July 14, 1831), when he headed a protest while wearing the uniform of the disbanded militia and heavily armed with a pistol, a rifle, and a dagger. Once again, he was arrested. In October, he was sentenced to six months in prison; then he was released on April 29, 1832.

We might say that Galois's time in prison was not completely wasted, since he continued to develop mathematical concepts there.[1] Another one of his papers was rejected while he was in jail, and he reacted violently and indicated that he would no longer publish through the academy, and only with his friend Auguste Chevalier. His rejection letter indicated that he needed to be more precise and less incomprehensible. He did take this advice and began to collect his mathematical manuscripts to bring them into a better intelligible fashion.

Now aged twenty, he was thrown into a pistol duel. There are numerous speculations as to how he was drawn into a duel against a person, who was

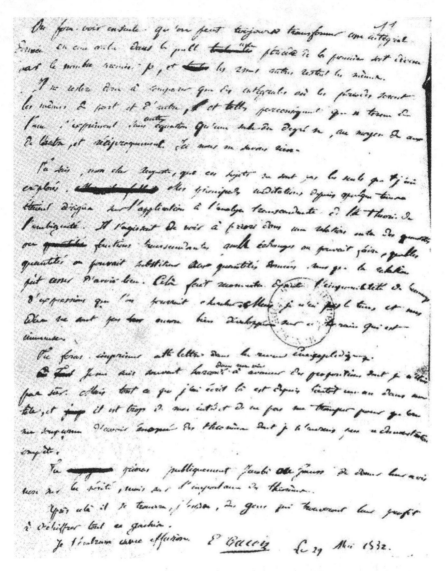

Figure 32.2. The final page of Galois's mathematical work, written on the eve of his death, has as its next to the last line, "*déchiffrer tout ce gâchis*" ("to decipher all this mess"). (Letter from Évariste Galois to his friend Auguste Chevalier, May 29, 1832.)

seen as a skilled marksman. There is great controversy as to what led to his fatal duel on May 30, 1832. Was it over competition for a woman's love? Was the opponent in the duel the woman's uncle or fiancé? Or was this duel simply staged by the police to eliminate a political enemy? In any case, Galois stayed up all night before the scheduled duel, writing letters to friends and attaching a manuscript of his works in a somewhat more intelligible form. It is believed that he suspected he would not be victorious in the upcoming duel. It is often believed that the material he left behind on that fateful night is today the basis of what we refer to as Galois theory.

The duel took place early in the morning of May 30, 1832.[2] He was shot in the abdomen and died the next morning in the hospital to which he had been taken by a passing farmer. His radical behavior did not end even at this sad time, as he refused the support of a priest and told his younger brother, Alfred: "*Ne pleure pas, Alfred! J'ai besoin de tout mon courage pour mourir à vingt ans!*" ("Don't cry, Alfred! I need all my courage to die at twenty!"). On June 2, Évariste Galois was buried in a common grave at the Montparnasse Cemetery.

CHAPTER 33

James Joseph Sylvester: English (1814–1897)

The British mathematician James Joseph Sylvester was born in London on September 3, 1814, and is known for having made significant contributions to combinatorics, matrix theory, number theory, and other branches of mathematics—while living in both the United States and England.[1] Curiously, the mathematician known as James Joseph Sylvester was not born by that name; because his father's name was Abraham Joseph, he was born as James Joseph. At the time that James's older brother came to the United States, it was required that all immigrants have a middle name as well as a surname. The older brother adopted Sylvester as his new surname, and the younger brother, James Joseph, did the same. At the early age of fourteen, he studied at the University of London with the famous English mathematician Augustus De Morgan (1806–1871). His family withdrew him from the university after he had a skirmish with another student. He then entered the Liverpool Royal Institution. His more serious study of mathematics continued in 1831 at St. John's College, Cambridge University. After several years of illness, he finally sat for the famous Cambridge mathematics examination, the Mathematical Tripos; he scored very high, coming in second in the competition. He was qualified to receive his university degree but did not receive it. Because doing so would conflict with his Jewish faith, he had refused to accept the Thirty-Nine Articles of the Church of England, which resulted in him being denied his degree. Furthermore, this also prevented him from obtaining the Smith's Prize or a subsequent fellowship.

Nevertheless, he became a professor of natural philosophy at the University College London in 1838, and the following year became a fellow of the Royal Society of London. In 1841, he was finally awarded a bachelor of arts degree and a master of arts degree from Trinity College in Dublin, Ireland. Shortly thereafter, he moved to the United States and became professor of mathematics at the University of Virginia. After four months, once again, a violent encounter with two students caused him to leave. Subsequently, he moved to New York City, where he was denied an appointment as professor of mathematics at Columbia University because of his Jewish religion. As a result, in November 1843 he left New York City for England.

Once back in England, he took on a leadership position at the Equity and Law Life Assurance Society, where he used his mathematical talents to develop actuarial models. However, this position required a law degree, so, subsequently, he studied for the bar examination. At this time, he met Arthur Cayley (1821–1895), another mathematician who was also studying law. He collaborated with Cayley for many years, where together they

Figure 33.1. James Joseph Sylvester,
sometime after his 1884 arrival in Oxford.

made significant contributions to matrix theory and invariant theory. It was
not until 1855 that Sylvester was once again appointed professor of mathematics, this time at the Royal Military Academy in Woolwich, in southeast
London. He stayed there until 1869, when he was forced to retire at age
fifty-five. He had to battle to receive a full pension, which he eventually
achieved. It was not until 1872 that Cambridge University finally awarded Sylvester's long overdue bachelor's and master's degrees, overcoming his
initial blockage on account of being Jewish.

In 1876, he returned to the United States at the invitation of the new
Johns Hopkins University in Baltimore, Maryland, where he became one
of its first professors of mathematics. While there, in 1878, he founded the
American Journal of Mathematics, which at that point was only the second
such professional publication available in the United States. He returned
to England in 1883 to accept the Savilian Professor of Geometry position
at Oxford University. Here, too, he was not very popular with students;
Sylvester tended to lecture primarily on his own research and was not too
concerned about spreading other mathematical knowledge to his students.
In time, his capacities weakened, including memory loss and poor eye-
sight; in 1892, although he retained his position at Oxford, he returned to
London and spent his last years at the Athenaeum club, until his death on
March 15, 1897.

James Joseph Sylvester is remembered for a number of mathematical
developments, as well as the terms he introduced to our mathematics language, such as *matrix*, *discriminant*, and *graph* in the field of combinatorics. In fact, "he once laid claim to the appellation 'Mathematical Adam,'
asserting that he believed he had 'given more names (passed into general
circulation) to the creatures of the mathematical reason than all the other
mathematicians of the age combined.'"[2] He also came up with an interesting
way of developing the value of π:

$$\pi = 2 + \cfrac{2}{1 + \cfrac{1 \cdot 2}{1 + \cfrac{2 \cdot 3}{1 + \cfrac{3 \cdot 4}{1 + \cfrac{4 \cdot 5}{1 + \cfrac{5 \cdot 6}{1 + \cdots}}}}}}.$$

Sylvester had a deep knowledge of classical literature, and he peppered
his mathematical papers with Latin and Greek quotations. If we were to

categorize Sylvester, we would probably say that he was largely an algebra-ist. He did some outstanding work in number theory, where, for example, he showed the number of possible ways a number can be expressed as a sum of positive integers. He worked with Diophantine equations, which are algebraic equations that require integer solutions. He also liked related problems, such as "I have a large number of stamps to the value of only 5*d* and 17*d*. What is the largest denomination, which I *cannot* make up with the combination of these two different values?" (The correct answer is 63*d*.)

Not only did he enjoy presenting puzzles to a general audience as well as to mathematicians, but he also took pride in being able to compose poet-ry. Moreover, he had a great interest in music and even took singing lessons from Charles Gounod (1818–1893). Sylvester wrote: "May not music be described as the mathematics of the sense, mathematics as music of the reason? The musician feels mathematics, the mathematician thinks music: music the dream, mathematics the working life."[3] Sylvester leaves a rather broad legacy.

CHAPTER 34

~

Ada Lovelace:
English (1815–1852)

In this technological age, when the computer pretty much guides us through our daily lives, the profession of computer programmer has become popular as a result. A curious person might ask: Who was the first computer programmer? Consensus has it that English mathematician Augusta Ada King Noel, Countess of Lovelace—more commonly known as Ada Lovelace—holds that honor.

It all began when Lovelace was seventeen and scientist Mary Somerville introduced her to the mathematician Charles Babbage. Babbage showed her his just-developed invention, the Difference Machine, which is considered the world's first mechanical calculator (see chap. 30). Babbage also conceived of the Analytical Engine, which had been intended to do much more than mere subtractions; unfortunately, it was never actually built. Babbage, who was twenty-four years her senior, was enchanted with Lovelace's interest in mathematics and science. They began a twenty-year correspondence. Lovelace remained enthusiastic for mathematics throughout her life. For example, at age twenty-five, she contacted the well-known British mathematician and first professor of mathematics at the University of London, Augustus De Morgan, and asked him to tutor her in mathematics. At one point, De Morgan wrote to Lovelace's mother, indicating that Ada Lovelace had an extraordinary talent in mathematics, which would have made her quite famous, if she were a man.

Figure 34.1. Ada Lovelace, portrait ca. 1840.

In 1843, Lovelace produced what we consider today the first foray into computer programming. The story begins in 1841, when Babbage was invited to give a lecture at the University of Turin to describe his Analytical Engine. Luigi Menabrea, a mathematician who would eventually become prime minister of Italy, took notes on the lecture and transcribed them in French. In 1843, Babbage's friend Charles Wheatstone asked Lovelace to translate the French notes into English as she was fluent in French. She not only translated the work but also added her own notes about the lecture. One such addition was her description of an algorithm for the Analytical Engine that could compute the Bernoulli numbers (a sequence of rational numbers that occur frequently in number theory—see chap. 22). In so doing, she became the first person to write an algorithm for a machine to produce more than just a simple calculation. Lovelace earned the honor of first computer programmer in the history of mathematics with her achievement. In figure 34.2, we show a diagram contained in Lovelace's notes—which, by the way, were far more voluminous than the mere translation requested of her.

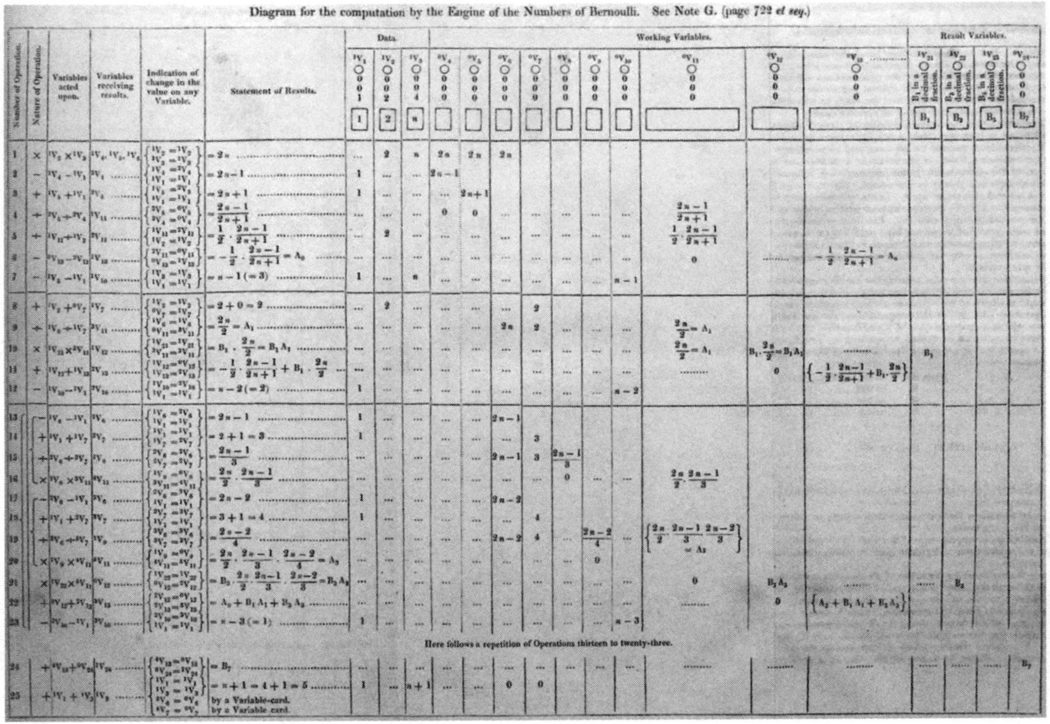

Figure 34.2. Lovelace's diagram of her algorithm for the Analytical Engine to compute the Bernoulli numbers, which she included with her translation. (From *Sketch of the Analytical Engine Invented by Charles Babbage* by Luigi Menabrea, with notes by Ada Lovelace [London: Richard and John E. Taylor, 1843].)

Now that we are familiar with her mathematical achievements, let's examine the life of Augusta Ada King Noel, Countess of Lovelace. She was born on December 10, 1815, in London, England, to her parents, Lady Byron (Anne Isabella Noel Byron, 11th Baroness Wentworth and Baroness Byron, nicknamed Annabella) and the famous British poet Lord Byron (George Gordon Byron, 6th Baron Byron). Unfortunately, a month after Ada's birth, Lord Byron separated from his wife; several months later, he left England forever. In the third canto of *Childe Harold's Pilgrimage: Harold the Wanderer*, Lord Byron mentioned his daughter: "Is thy face like thy mother's my Fair child! ADA! sole daughter of my house and heart?"[1] Byron died in 1824 in Missolonghi, Greece.

Lovelace had an unusual early life. Her mother was truly angry at the departure of her husband, so Ada grew up never having seen even a picture of her father. That did not happen until she was twenty years old. Lovelace was largely reared by her maternal grandmother, Judith Milbanke, and she suffered a number of childhood illnesses. For example, in 1829, she spent nearly a year in bed, suffering from a paralysis that evolved from a bout with the measles. Lovelace was interested in not only mathematics but also all things mechanical and scientific. For example, she was fascinated by the notion of flying, which prompted her to write a book titled *Flyology*, even though she was still just an adolescent. Her work in *Flyology* illustrated her understanding of what would be required for humans to fly like birds, considering, for instance, the size of wings that humans might need to use in order to fly. Motivated by her interest in science, Lovelace sought out many of the top scientists in England. Notable among these was Michael Faraday, who made major advances in electromagnetism.

In 1834, Lovelace began to attend regular court events, where she charmed people with her intelligence and her dancing talent. On July 8, 1835, she married William, 8th Baron King; as Lady King, Lovelace entered into a rather wealthy environment. Over the next four years, she gave birth to three children, Byron, Anne Isabela, and Ralph Gordon. Since she was a descendent of Baron Lovelace, in 1838, her husband became the Earl of Lovelace, and she became the Countess of Lovelace. Her mother continued to stay involved with the family; she hired tutors to support the three children and ensured that her daughter remained morally correct.

Perhaps it was Ada Lovelace's interest in mathematics, which spurred her later love of gambling. In the late 1840s, betting on horses resulted in

Figure 34.3. Portrait of
Ada Lovelace.
(Painting by Margaret
Sarah Carpenter, oil on
canvas, 1836).

her loss of over £3,000. In 1851, she tried to develop a mathematical model to guide her to successful bets, but this was a total financial disaster. Despite these struggles, Lovelace is lauded to this day for her insight into the potential of Babbage's Analytical Engine to bring further developments in mathematics beyond merely doing arithmetic calculations. In her notes to the translation of Babbage's lecture, she included the following:

Again, it [the Analytical Engine] might act upon other things besides number, were objects found whose mutual fundamental relations could be expressed by those of the abstract science of operations, and which should be also susceptible of adaptations to the action of the operating notation and mechanism of the engine. . . . Supposing, for instance, that the fundamental relations of pitched sounds in the science of harmony and of musical composition were susceptible of such expression and adaptations, the engine might compose elaborate and scientific pieces of music of any degree of complexity or extent.

The distinctive characteristic of the Analytical Engine, and that which has rendered it possible to endow mechanism with such extensive faculties as bid fair to make this engine the executive right-hand of abstract algebra, is the introduction into it of the principle which Jacquard devised for regulating, by means of punched cards, the most complicated patterns in the fabrication of brocaded stuffs. It is in this that the distinction between the two engines lies. Nothing of the sort exists in the Difference Engine. We may say most aptly that the Analytical Engine weaves algebraic patterns just as the Jacquard-loom weaves flowers and leaves.[2]

These lines from her notes offer a good indication of how her vision reached far into the future.

On November 27, 1852, at the age of thirty-six, Lady Ada Lovelace died from uterine cancer. During her fatal illness, she was comforted and cared for by her mother, Annabella. One of the many famous people whom Lovelace met during her life was the famous author Charles Dickens. In August 1852, Dickens visited his bedridden friend and, at her request, read her a well-known scene from his 1848 novel, *Dombey and Son*, in which a six-year-old boy dies. As she wished, Lady Ada Lovelace was buried next to her father, Lord Byron, inside the church of St. Mary Magdalene in Hucknall, England.

Through the twentieth century, Lovelace has been remembered through books (*The Difference Engine*, by William Gibson and Bruce Sterling), plays (*Childe Byron*, by Romulus Linney), and films (*Conceiving Ada*, directed by Lynn Hershman Leeson). However, it should be noted that Lovelace's fame was brought into the fore in 1953, when her notes were republished in the book *Faster Than Thought: A Symposium on Digital Computing Machines*, by B. V. Bowden. Today, in the United Kingdom, the second Tuesday of October is designated as Lady Ada Lovelace Day. As recently as 1980, the

United States Department of Defense honored Lady Lovelace by naming a newly developed computer language with her first name, "Ada." Her legacy lives on as the first computer programmer, not necessarily the first woman computer programmer, but in actual fact, the first person to be a computer program. Moreover, she really was a visionary in that she realized the significance of Babbage's Analytical Machine, foreseeing the versatile applicability of programmable computers.

CHAPTER 35

~

George Boole:
English (1815–1864)

The English mathematician and logician George Boole developed a logical theory that serves today as a basis for electronic devices, and of course, the modern digital computer. In mathematical circles, Boolean algebra, which we will introduce later, has made his name popular to this day.

George Boole was born on November 2, 1815, in the town of Lincoln, Lincolnshire, England. Although his father was a shoemaker, he provided regular lessons for his son, which included making optical instruments. Besides a few years in elementary school, Boole was largely self-taught in mathematics. To help support the family, Boole had taught in local elementary schools at the age of sixteen, and at twenty he opened his own school. During his leisure time he read classical mathematical books by such famous mathematicians as Isaac Newton, Pierre-Simon Laplace, and Joseph-Louis Lagrange. He was taught Latin by some local folks, but he was self-taught with modern languages. He continued to be popular locally in matters of education and other social issues. However, throughout this time he continued to study mathematics and began to publish papers, especially in algebra, using symbolic methods.

In 1849, he was appointed professor of mathematics at Queens College, Cork, Ireland, where he met his wife, Mary Everest, who, in her own right, became a mathematician of note. This marriage produced five daughters. In 1854, Boole wrote a treatise on Aristotle's system of logic entitled *An Investigation of the Laws of Thought, on Which Are Founded the Mathematical*

Figure 35.1. George Boole.

Theories of Logic and Probabilities. This was the basis of what later became known as Boolean algebra, which was based on simply two quantities: true or false, or 1 or 0. No other symbols are used in Boolean algebra aside from 1 and 0.

Let's take a quick look at some of the basics of Boolean algebra. First, there is the addition using only the two symbols available, 1 and 0:

$$0 + 0 = 0$$
$$0 + 1 = 1$$
$$1 + 0 = 1$$
$$1 + 1 = 1$$

This is analogous to the "or" function in logic, where the 1 can replace "true" and the 0 can replace "false." That is, if either of the two elements being added is a 1, then the sum is 1.

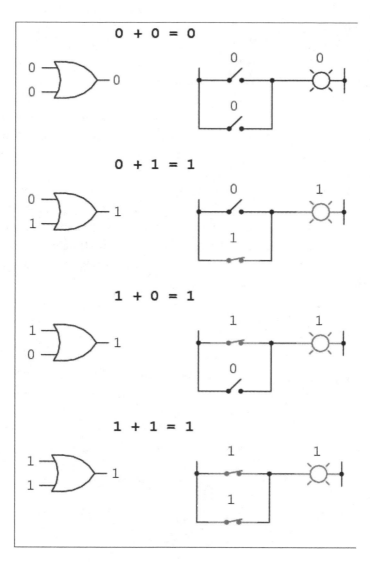

Figure 35.2.

So if we have a longer addition, the same holds true. Such as: 1 +0 + 1 + 1 +1 + 0 = 1. This can also be seen with switching circuits, as shown in figure 35.2.

In Boolean algebra we also have multiplication, which follows the "and" rules for logical reasoning. That is, something is true when *both* are true. This can be shown as follows symbolically:

$0 \times 0 = 0$
$0 \times 1 = 0$
$1 \times 0 = 0$
$1 \times 1 = 1$

We see here that in order for us to get a 1 or a true statement, both must be true; that is, a true *and* a true yields a true. Once again, we can see that with switching circuits to get the light to go on, both switches must be closed (see fig. 35.3).

In Boolean algebra, if a statement, or a variable, is not 1, then it is 0, so we can say that 0 is a complement of 1. We could go on and develop the entire Boolean algebra, but the size of this book would not be large enough to accommodate the subject. We leave this to the ambitious reader to pursue further. However, this system allowed logical arguments to obtain more structure.

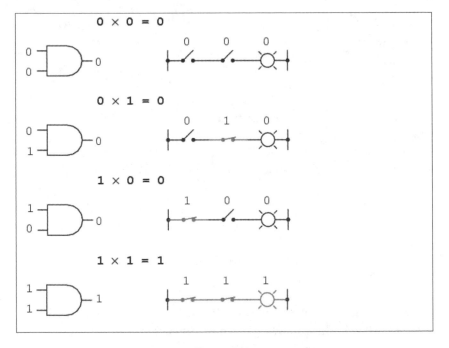

Figure 35.3.

George Boole continued to become more involved in social matters as well as university instruction. At the end of November 1864 on his way to the university from home, a mere distance of 3 miles, Boole got caught in a severe rainstorm and continued to lecture in his wet clothing. Shortly thereafter, he came down with pneumonia, which worsened, and on December 8, 1864, he died at Ballintemple, Cork, Ireland. Today, George Boole is largely remembered through his development of Boolean algebra and a crater on the moon named after him.

CHAPTER 36

~

Bernhard Riemann:
German (1826–1866)

In many European countries, a doctoral thesis is alone not sufficient to obtain a professorship at a university. In addition, one has to write a habilitation thesis that must be reviewed by and successfully defended before an academic committee. The process of a habilitation is often seen as a second doctoral dissertation. The habilitation is a postdoctoral qualification required to independently teach a subject in Europe at the university level. It must be accomplished without guidance of a supervisor, and it defines a higher level of scholarship than a doctoral dissertation. In mathematics, as well as in the natural sciences, the typical habilitation period is four to ten years. During this time, numerous research articles in high-quality scientific journals must be published. The thesis can then be either cumulative (that is, basically a collection of selected publications) or a monograph. If the thesis has been accepted, the applicant has to give a public lecture after which the habilitation is awarded. There is no concept of a habilitation in the United States, and there is an ongoing political discussion in Germany and other countries to abolish the system of the habilitation; it is a time-consuming obstacle in an academic career, contributing to the brain drain of promising young researchers to the United Kingdom and the United States, where their chances of getting a professorship at a reasonable age are often better.

In 1853, Bernhard Riemann was in the final stages of his habilitation at the University of Göttingen. He had been working for nearly three years

Figure 36.1. Bernhard Riemann.

on it, achieving several important new results and solving open problems related to the representation of functions by trigonometric series (these are infinite sums of sine and cosine functions with different wavelengths and amplitudes). He had also introduced a mathematically rigorous concept for the integral of discontinuous functions, a notion that would later be called the Riemann integral. Having already earned some reputation as a mathematician, the only missing step necessary to complete his habilitation was to give a public lecture before the habilitation committee. By the rules of the university, Riemann had to submit titles of three different lectures belonging to different areas of mathematics, from which the faculty of philosophy was to choose one. Riemann had already worked out the details of two of the three topics, but not for the third one, titled *"Über die Hypothesen, die der Geometrie zu Grunde liegen"* (On the hypotheses that lie on the foundations of geometry). Although, or perhaps because, it was the topic least related to Riemann's prior work and interests, it was chosen by the faculty, more precisely by Carl Friedrich Gauss, professor at Göttingen and also Riemann's doctoral thesis supervisor. Riemann's habilitation lecture, on which he worked for several months, became a famous classic of mathematics, changing the whole subject of geometry. By introducing completely new and brilliant viewpoints on geometrical ideas, he was able to generalize geometrical concepts and even the notion of space itself in a way that would evoke a whole new branch of mathematics. Before we expose some of Riemann's groundbreaking ideas and its consequences, we want to give a brief overview of his life.

Georg Friedrich Bernhard Riemann was born on September 17, 1826, in Breselenz, a village in the Kingdom of Hanover (now Germany). His father was a poor Lutheran pastor; his mother died before her children had reached adulthood. Until the age of fourteen, Bernhard was educated at home by his father, assisted by a teacher from a local school. He was a shy and anxious child. In 1840, Bernhard went to the city of Hanover to live with his grandmother and attend the Lyceum (middle school), where he immediately entered the third-year class. Two years later, his grandmother died, and Bernhard moved to Lüneburg to continue his education at the Johanneum Gymnasium (high school). While he was not very good in languages, history, or geography, he showed exceptional skills and interest in mathematics. His teachers soon recognized his incredible talent, and the school principal allowed him to study mathematics books from his own library, among them Legendre's 900-page book on the theory of numbers, which Bernhard read in six days. In 1846, Riemann enrolled at the University of Göttingen to study theology and become a pastor like his father. However, his strong interest in mathematics naturally made him attend some mathematics lectures as well. It became clear to him that what he really wanted to study was mathematics. He asked his father for permission and began to take courses in mathematics as a regular student. Among his teachers in the elementary courses were well-known mathematicians Moritz Stern (1807–1894) and Johann Benedict Listing (1808–1882). Gauss, the most famous mathematician at Göttingen, was mainly teaching astronomy at that time. After one year in Göttingen, Riemann moved to Berlin to study advanced topics under Peter Gustav Dirichlet (1805–1859), Gotthold Eisenstein (1823–1852), Carl Jacobi (1804–1851), and Jakob Steiner (1796–1863). Of his teachers in Berlin, Dirichlet probably had the greatest influence on Riemann. Dirichlet always tried to condense the essence of a mathematical theory into an intuitively comprehensible idea and then use this idea as a guiding principle to find new mathematical results. Riemann embraced Dirichlet's style of doing mathematics and he was full of ideas when he returned to Göttingen in 1849. He began to write his doctoral dissertation under the supervision of Gauss and got a temporary position as the assistant to the physicist Wilhelm Weber (1804–1891), from whom he learned a lot about theoretical physics. His doctoral thesis was completed in 1851 and in his report on the thesis, Gauss describes Riemann as having "a gloriously fertile originality." With Gauss as a mentor, Riemann started to work on his habilitation.

When Riemann finally delivered his habilitation lecture "On the hypotheses which underlie geometry," only Gauss, one of the leading geometers at the time, was able to fully recognize the significance of Riemann's work and was deeply impressed by it. The lecture contained hardly any formulas; it was not a purely mathematical presentation but rather a philosophical treatise about the meaning of geometrical concepts, thereby identifying certain implicit hypotheses on the nature of space on which our understanding of geometry is based. The ingenuity of Riemann's thoughts was to give up these hypotheses and replace geometrical notions relying on them by more general concepts that can be formulated without any pre-assumptions on the underlying space. What are these hypotheses that Riemann abandoned?

Well, you may recall that the geometry taught in high school is called Euclidean geometry; it is based on five postulates attributed to the Alexandrian Greek mathematician Euclid. The fifth of his postulates essentially says that two parallel lines never cross, even if we extend them infinitely. This, however, is just an assumption that cannot be proved or verified by experiment. To illustrate that the parallel postulate is by no means trivial, let's consider the surface of the Earth, which, for simplicity, we may think of as a perfect sphere. We know that if we draw two lines on a sphere, starting out parallel or "in the same direction," they will inevitably meet at some point. Think of two longitudinal circles that define perfectly parallel tracks close to the equator, but they converge at the poles. Now you may object that lines on a sphere are not actually straight, so the parallel postulate doesn't apply here. We know that the Earth is not flat and looking at our planet from some distance, it's obvious that the lines of fixed longitude are not straight lines in space. But imagine for a moment that we were human beings living not on the surface of our planet, but "in the surface," meaning that we were two-dimensional beings living in a two-dimensional world as in the famous satirical novella *Flatland: A Romance of Many Dimensions* by the English schoolmaster Edwin A. Abbott, first published in 1884. How would a "flatlander," confined to the two dimensions of a sphere, without the possibility

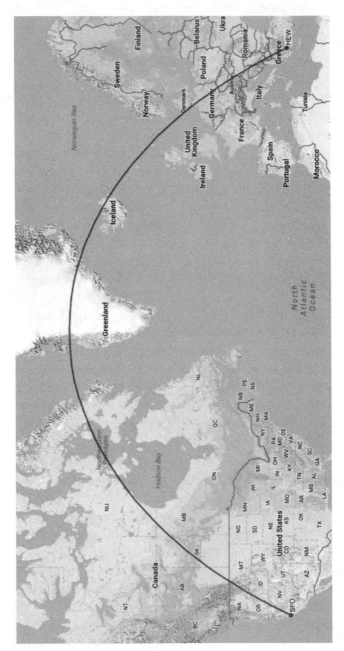

Figure 36.2.

to move in a third dimension, define a line to be straight? Well, a straight line is the shortest connection between any two given points. But the shortest connections between points on a sphere are indeed arcs of great circles (circles on the surface of the sphere and with their center at the center of the sphere)—they are the shortest lines connecting two points on a sphere. This is also the reason that airplanes flying from the West Coast of the United States to Europe will fly over Greenland; they follow a great circle path. Figure 36.2 shows the arc of the great circle connecting San Francisco and Athens, Greece. The two cities have almost the same latitude, but flying eastward along a circle of latitude would amount to a much longer distance than flying northward to Greenland and then southward to Athens. If you have a globe handy, this is easy to see. Only on a flat map, where proportions are distorted, the "straight" route along a circle of latitude appears to be shorter.

For "flatlanders" living on the surface a sphere, a great circular arc would appear "as straight as straight can be," since their world is made up of only two dimensions, and they cannot see that their two-dimensional space is actually curved. What an observer in the surrounding three-dimensional space would call a great circle would appear to be a straight line for them. Moreover, if their habitat would only cover a very small patch of the sphere (think of human-sized creatures on the surface of the Earth), they would never find out that these perfectly straight lines would eventually meet, if only extended long enough. If we let a ball roll freely on the surface of the Earth, it would also trace out a great circle, but locally its path will look like a straight line. Two balls starting out side by side and rolling freely with equal velocity would eventually collide, since they would follow great circle routes. However, because we are so much smaller than the Earth, we are not able to detect the curvature of the Earth from the observation of balls rolling on its surface. There are other effects from which the spherical shape of the Earth can be deduced. But how could flatlanders find out whether they are living on a flat plane or on a curved surface like a sphere? Apart from measuring whether the distance between two straight lines remains constant as the lines are extended, they could also measure the sum of the angles in a triangle. You know that the sum of the angles of a triangle on a plane is always 180 degrees. But this is not true for triangles on a sphere! The sum of the angles of a triangle on the sphere is always greater than 180 degrees; actually, the sum of the angles of a spherical triangle is between 180 degrees and 540 degrees. For example, the triangle ABC shown in figure

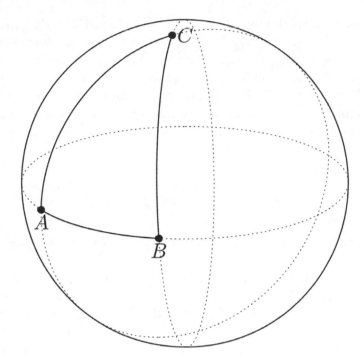

Figure 36.3.

36.3 has two right angles at the vertices *A* and *B*, already yielding 180 degrees together, plus the angle at vertex *C*.

To see that the angle sum can get arbitrarily close to 540 degrees, keep points *A* and *C* fixed and move point *B* eastward along the equator, around the sphere until it almost touches point *A*. Then we still have right angles at *A* and *B*, but the angle at vertex *C* will get arbitrarily close to 360 degrees, and hence the sum of all three angles can be arbitrarily close to 540 degrees. Thus, we have convinced ourselves that on the surface of a sphere, it is not true that the sum of the angles of a triangle is 180 degrees, nor is it true that parallel lines never intersect. Both statements are rather obvious, if we look at the sphere as a two-dimensional surface in three-dimensional space; they can also be turned around—that is, if we find that the sum of the angles of a triangle is not 180 degrees, then the surface on which this triangle is drawn must be curved. This implies that creatures living in a two-dimensional space can, at least in principle, find out whether their space is curved. Riemann's ingenuity was to realize that it is not only possible to determine the curvature of a two-dimensional space without knowing about an ambient three-dimensional space (for instance by measuring the angles of a triangle), but an ambient higher-dimensional space doesn't even have to exist for a space to be curved. He invented the notion of n-dimensional manifolds to describe spaces in which the Euclidean parallel postulate is not necessarily true, and in which the sum of the angles of a triangle is larger or smaller than 180 degrees. Moreover, he introduced mathematical tools and objects that allow us to study properties of general manifolds, for instance curvature. For example, the geometric properties of the surface of a sphere can be described entirely in terms of a two-dimensional Riemannian manifold without referring to an ambient three-dimensional space. But what makes us so sure that the three-dimensional space we live in is not curved? Could it be that two parallel lines would eventually meet, perhaps after a distance of thousands of light years? Riemann already contemplated such possibilities in his habilitation lecture, viewing our three-dimensional world as a three-dimensional manifold that might be curved. Unfortunately, the brilliancy and the consequences of Riemann's work, forming the basis of what is now called Riemannian geometry, was not understood by his contemporaries; his ideas were too far ahead of his time.

After his habilitation, Riemann began to lecture at the university. At that time, the main part of a lecturer's income was tuition fees from the students. Riemann's income was particularly low, since he was teaching advanced courses that only a handful of students would attend. Yet he was still happy with his job and teaching helped him to slowly overcome his shyness, as some touching letters to his father reveal. In spite of his low budget, he supported two of his sisters, who were sent to him after his brother died. Riemann's mentor, Gauss, died in 1855, but Riemann, still in his late twenties, unfortunately did not get Gauss' chair at Göttingen University. Dirichlet, twenty years older than Riemann, became the successor to Gauss and attempts to create a permanent position for Riemann failed. However, in 1857, the university at least granted him a regular salary. Riemann was an exceptionally versatile mathematician, interested in various branches of mathematics as well as physics. His works were always guided by intuitive ideas that he also tried to convey and employ in his publications, often allowing him to avoid tedious computations that would not provide any deeper insight. This is also one of the reasons that mathematicians still find his publications very inspiring to read. They are by no means outdated or old-fashioned! In 1859, Dirichlet died and Riemann was finally appointed to the chair of mathematics at Göttingen University and also elected to the Berlin Academy of Sciences. Financially secure, he married Elise Koch in 1862. In the autumn of that year, Riemann got a severe cold that developed into tuberculosis. He had had problems with his health throughout his life, and his mother as well as four of his six siblings had died young. Hoping to improve his health in a warmer climate, Riemann went to Sicily and stayed there during the winter. He returned to Göttingen University, but his physical condition soon got worse again and he moved back to Italy, and this time he stayed for over one year. Following another period at Göttingen University, Riemann died on this third journey to Italy in Selesca on Lago Maggiore on July 20, 1866.

In 1905, Albert Einstein published his special theory of relativity, in which space and time are no longer independent, but combined into a four-dimensional space-time, or, as Hermann Minkowski (1864–1909) wrote:

> Space by itself, and time by itself, are doomed to fade away into mere shadows, and only a kind of union of the two will preserve an independent reality.

After this revolutionary milestone of physics, Einstein began to search for a way to incorporate gravity into his theory. During his attempts he came across Riemann's theory of manifolds to describe curved spaces, which would turn out to be the key to a relativistic theory of gravity. However, Einstein was not a mathematician, so he needed help. The mathematician Marcel Grossmann (1878–1936), a friend and former colleague, mentored Einstein in Riemannian geometry and collaborated with him. After ten years of "blood, sweat, and tears," Einstein presented his general theory of relativity, which many consider to be the most beautiful physical theory ever invented. In it, gravity is described as the curvature of four-dimensional space-time, which is represented by a so-called semi-Riemannian manifold. Einstein's field equations specify how matter and radiation influence the geometry of this four-dimensional manifold by relating the Riemann curvature tensor to the distribution of matter and radiation. Riemann's theory of curved spaces was the key ingredient in Einstein's theory of general relativity, and it is remarkable that in his habilitation lecture, Riemann was already "on the right track" when he pondered the possibility that the physical space we live in could actually be a curved space.

CHAPTER 37

∿

Georg Cantor:
German (1845–1918)

William Shakespeare is without a doubt one of the greatest writers and dramatists in history. Many consider him to be the greatest, at least in the English language. His plays are still very popular throughout the world, although they were written about 400 years ago. They are timeless master-pieces and they will never become out of date, since they can easily be rein-terpreted in diverse cultural and political contexts in modern productions. While there exist immense amounts of literature on Shakespeare's works, little is known about his life. In fact, the biographical records are so sparse that in the middle of the nineteenth century, doubts regarding the author-ship of the works attributed to him began to be expressed. Proposed al-ternative candidates include the philosopher and statesman Francis Bacon (1561–1626), the poet and playwright Christopher Marlowe (1564–1593), and Edward de Vere, 17th Earl of Oxford (1550–1604). Today, only a very small minority of academics still pursue theories on alternative author-ships, which are, however, generally considered as fringe beliefs, against the scholarly consensus that William Shakespeare is indeed the author of the works published under his name. Yet during the last decades of the nine-teenth century, discussing theories supporting or refuting the existence of a hidden author of the Shakespearean works was very fashionable in various academic circles, not only among the experts in the field.

In 1896, and the following year, the well-known German mathemati-cian Georg Cantor (1845–1918) published two pamphlets making a case for

Figure 37.1. Georg Cantor.

the theory that Francis Bacon was Shakespeare, which remained the most popular of the alternative-authorship theories until the early twentieth century. Cantor had begun an intense study of Elizabethan literature in order to distract him from mathematics after he had suffered a severe personal crisis. His crisis was triggered by the strong criticism and rejection of his mathematical work by some of the most distinguished mathematicians of his time. The French mathematician Henri Poincaré (1854–1912) referred to Cantor's ideas as a "grave disease" infecting the discipline of mathematics, and the German mathematician Leopold Kronecker (1823–1891) attacked Cantor even personally, calling him a "scientific charlatan" and a "corrupter of youth." What was so controversial about Cantor's work?

Georg Cantor was born in St. Petersburg, Russia, on March 3, 1845, and moved with his family to Germany at age 11; throughout his school years he showed outstanding skills in mathematics and in 1860 he graduated from high school with distinction. He then studied mathematics at the Swiss Federal Polytechnic and at the University of Berlin, receiving his doctorate degree in 1867. Cantor, often referred to as the originator of set theory, had invented new notions of mathematical infinities, based on a brilliant idea for comparing sets containing infinitely many elements. An example of a set containing infinitely many elements is the set of the natural numbers $\mathbb{N} = \{1, 2, 3, ...\}$, but there also exist other infinite sets of numbers. So, let's see what happens when we try to compare two infinite sets. Let us consider the set of all integers $\mathbb{Z} = \{..., -3, -2, -1, 0, 1, 2, 3, ...\}$ and compare its size with the set $\mathbb{N}$. We may conclude that $\mathbb{Z}$ is substantially larger than $\mathbb{N}$, essentially twice as large, to be more specific. This seems to be a perfectly reasonable and modest assumption that nobody would question. But since we know that $\mathbb{N}$ is infinite, our assumption of $\mathbb{Z}$ being larger than $\mathbb{N}$ implies that $\mathbb{Z}$ should be even "of greater infinity," in some sense. But what should that mean? Moreover, between any two integers we can find numerous fractions. Hence the set of all fractions (or rational numbers), $\mathbb{Q}$, should be "of much greater infinity" than $\mathbb{Z}$, but how much more? And what about all the real numbers? Is there any way to "measure" infinities? Cantor was the first to discover that such questions can indeed be addressed and answered in a mathematically rigorous way. He found a very simple, but brilliant, method to compare the sizes of different sets, even if they are infinite. He also established the basic mathematical notion of a set and developed the field of *set theory*, which is now one of the foundations of modern mathematics. To explain Cantor's brilliant idea for comparing sets, suppose we are given two sets, *A* and *B*, both of which contain only a finite number of elements (see fig. 37.2).

Then one (and only one) of the following three statements must be true:

1. Set *A* has more elements than set *B*.
2. Set *A* has fewer elements than set *B*.
3. Both sets *A* and *B* contain the same number of elements.

Is there any way to find out which of these statements is true without actually counting the elements of *A* and *B*? Yes, there is! We just have to pair each member of *A* with a corresponding member of *B*, for instance, by drawing a line from one to the other (see fig. 37.3).

If we manage to do this for all elements of *A* and *B* and no elements are omitted from either set, then for each member of *A* there must be exactly one "partner" element in *B*, and thus both sets must contain the same number of elements. In mathematics, this is called a *one-to-one correspondence*

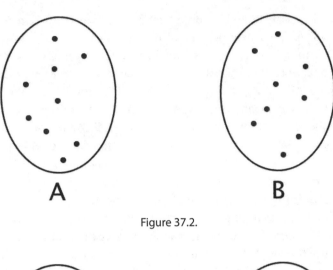

Figure 37.2.

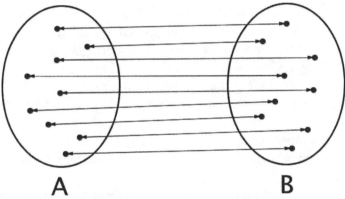

Figure 37.3.

between the elements of the two sets. Although this method of comparing sets is, in fact, very old, since it is actually nothing other than "counting with fingers," Cantor was the first to recognize that this strategy can also be applied to infinite sets.

Cantor's notion of a one-to-one correspondence enables us to compare two infinite sets, since we don't have to actually count the number of elements in each of the sets separately, and then compare the numbers. We just need to find out whether we can establish a one-to-one correspondence between the elements of the two sets. Above we tried to convince you that there are many more rational numbers (or fractions) than integers and more integers than natural numbers. Astonishingly, this is wrong. It is possible to establish a one-to-one correspondence between $\mathbb{N}$ and $\mathbb{Z}$, for example by numbering the members of $\mathbb{Z}$ as follows: Let the first integer in our sequence be the number 0, the second integer the number 1, the third integer -1, the fourth 2, the fifth -2, the sixth 3, and so on. This numbering scheme obviously puts $\mathbb{N}$ and $\mathbb{Z}$ into a one-to-one correspondence, showing that both sets are actually "of equal size," which is counterintuitive and upsetting, even for many mathematicians at the time Cantor published his ideas. Cantor even showed that also the rational numbers can be put into a one-to-one correspondence with the natural numbers. Another way of saying this is that we can give all the rational numbers a waiting number in an infinite line. We won't give a detailed proof here, but the essential idea in Cantor's proof is not too difficult to grasp. Considering only positive fractions for the moment, we can order them in a table by placing the fraction $\frac{p}{q}$ in the cell at the intersection of row p and column q (see fig. 37.4).

Table of all fractions:

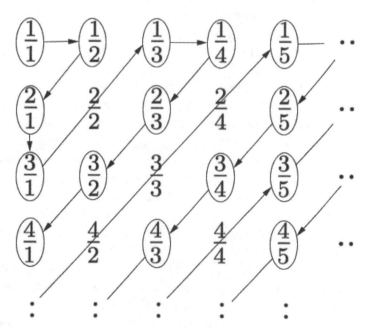

Figure 37.4.

For instance, the fraction $\frac{73}{111}$ will be in the table at the intersection of the 73rd row and the 111th column. Now we want to put all positive fractions in a waiting line. Of course, this waiting line will never end, since the table never ends either; but that doesn't matter. We only have to make sure that every fraction will be included. To achieve this, Cantor proposed a clever "diagonal" counting scheme: We start at $\frac{1}{1} = 1$ and draw an arrow to the right, getting to $\frac{1}{2}$. From here we move on diagonally downward to $\frac{2}{1} = 2$, then straight downward to $\frac{3}{1} = 3$, then diagonally upward, arriving at $\frac{1}{3}$ (we skipped $\frac{2}{2} = 1$ because it has already been counted). Now the whole procedure is repeated—that is, "one to the right and diagonally downward until we reach the first column, then straight down and diagonally upward." Whenever we encounter a fraction that is equivalent to one that has already gotten a number, we skip it (these are the bypassed fractions in fig. 37.4). To see why a diagonal counting scheme is essential, the following illustration might be helpful: Suppose you have a robotic lawnmower and the infinite table of fractions in figure 37.4 defines the area to be mowed. How should the lawnmower move in order to reach every piece of this infinite lawn? Since the infinite lawn has only one corner, it must start there and work its way in diagonal serpentines away from that corner—following the infinite waiting line drawn in figure 37.4. By using Cantor's clever diagonal scheme, we manage to put all fractions in a waiting line without omitting any one of them. We have, therefore, established a one-to-one correspondence between all positive fractions and the natural numbers: The first fraction is $\frac{1}{1}$, the second fraction is $\frac{1}{2}$, the third one is $\frac{2}{1} = 2$, the fourth one is $\frac{3}{1} = 3$, and so on (see fig, 37.4). Every fraction gets a number given by its position in the waiting line. So far we have omitted the negative fractions, but we can now simply slip each negative fraction after the corresponding positive one in the line, and place the zero at the very first position. Because we can pair each rational number with a natural number, and no numbers are omitted from either set, there must be just as many natural numbers as there are rational numbers. A set whose members can be put in a waiting line without missing any one of its elements is called a countable set. So Cantor proved that $\mathbb{Q}$ is countable, a very surprising result!

Encouraged by this result, we might ask whether it is also possible to put the real numbers (that is, the rational and the irrational numbers) into a one-to-one correspondence with the natural numbers. Cantor showed that this is impossible, since no matter how cleverly we try to arrange the real numbers to put them in a waiting line, there will always be some numbers left over. To be precise, for any proposed list or counting procedure of all real numbers, we can always construct a number that cannot be included in this list. Real numbers can have infinite and nonrepeating sequences after the decimal point. It is this property that makes them "uncountable." An uncountable set is a set that contains too many elements to be countable and is, therefore, "larger" than the set of natural numbers, $\mathbb{N}$.

Suppose somebody claims to have found a procedure to enumerate all positive real numbers smaller than 1 (this is only a subset of all real numbers). All such numbers will start with a zero and a decimal point, followed by an infinite sequence of digits, if we append an infinite sequence of zeros whenever we encounter a number with a finite fractional part. You may ask him to tell you which number is first in his list, which number is second, and so on. Assume hypothetically that you write down each number, one after the other, thereby producing an infinite list of numbers with infinite fractional parts. We show that we can always write down a number between 0 and 1 that is not a member of his list, implying that his counting scheme does not include all real numbers. Our "magic number" must of course start with a zero and a decimal point. We obtain the first digit after the decimal point by looking up the first digit after the decimal point of the first number in his list and then write down the next-largest digit (for example, if we encounter 0, we write down 1, if we encounter 1, we write down 2, etc.) or, if the digit is 9, then we write down a 0. This will be the first digit after the decimal point of our "magic" number. As the next digit, we take the second digit after the decimal point of the second number in the list, change it according to our replacement scheme, and so on. By construction, our magic number differs from all numbers in the list, because it was constructed in a manner that its digit at position n does not coincide with the corresponding digit of the nth number in the list. It must be different from the first number in the list, because its first digit after the decimal point is different. It must also be different from the second number in the list, because its second digit after the decimal point is different. It must be different from the third number in the list, because its third digit after the decimal point is different, and so on. Therefore, this number cannot occur in the enumeration! This proof is now known as Cantor's diagonal argument, and the construction of a "diagonal sequence" from an infinite set of sequences became an important technique that is frequently used in mathematical proofs.

Cantor showed that it is impossible to enumerate the real numbers; they cannot be put into a one-to-one correspondence with the natural numbers. Hence, they are more "numerous" than the natural numbers, and thus, represent a "greater" infinity. He called such sets uncountable sets. Cantor also showed that among uncountable sets, there exist infinities of different "sizes" and he developed an arithmetic of infinities. To measure the size of infinite sets, he extended the natural numbers by numbers called "cardinals" and denoted them by the Hebrew letter $\aleph$ (aleph) with a natural number as a subscript. For instance, $\aleph_0$ (aleph-null) is the "cardinality" of the set of natural numbers—it is the "smallest" infinity in mathematics. At the time Cantor published these results, they shocked the mathematics community. They conflicted with common beliefs and were considered to be revolutionary. Cantor was well aware of the opposition his ideas were encountering. In his 1883 paper he wrote:

> I realize that in this undertaking I place myself in a certain opposition to views widely held concerning the mathematical infinite and to opinions frequently defended on the nature of numbers.

Many renowned mathematicians tried to prove Cantor wrong and did not accept his work. The criticism of his work threw him into a deep depression, and he even gave up mathematics for some time. He wrote:

> I don't know when I shall return to the continuation of my scientific work. At the moment I can do absolutely nothing with it, and limit myself to the most necessary duty of my lectures; how much happier I would be to be scientifically active, if only I had the necessary mental freshness.

At that time, Cantor took up his studies on Elizabethan literature to draw his own conclusions about the question of the authorship of the plays attributed to Shakespeare. Although his pamphlets on the subject did not stand the test of time, occupying himself with problems not at all related to mathematics probably helped him to recover from his depression. However, he never fully gained his passion for mathematics again. Instead he continued his research on a hidden author of the works of Shakespeare and published on his Bacon-Shakespeare theory until the end of his life. It took decades until the importance and ingenuity of his mathematical ideas were fully recognized. Cantor was ahead of his time. He had shown that the set of rational numbers, $\mathbb{Q}$, is not larger than the set of natural numbers, $\mathbb{N}$, which is a totally counterintuitive fact. Although this statement seems to contradict common sense, its proof is actually rather simple and not very difficult to follow. The same is true for the proof of the set of real numbers, $\mathbb{R}$, being essentially larger than $\mathbb{N}$, showing that there exist mathematical infinites of different sizes. That these very surprising and unexpected results can be found among the most innocent structures such as the natural, rational, and real numbers is one facet of the beauty of mathematics.

As one of the most distinguished foreign scholars, Cantor was invited in 1911 to attend the 500th anniversary of the founding of the University of St Andrews in Scotland. He was largely motivated to go so that he could meet Bertrand Russell, who had recently published his book *Principia Mathematica*, where he frequently cited Cantor's work. Unfortunately, that meeting never came to pass. In 1912, he was awarded an honorary doctorate from the University of St Andrews, but due to illness he was unable to accept the degree in person. Cantor retired in 1913, living in poverty and in ill health. In 1917 he was living in a sanatorium in Halle, Germany, much against his will and continuously asking to be released. Cantor's last years were plagued with illness and he died at the sanatorium of a heart attack on January 6, 1918.

CHAPTER 38

~

Sofia Kovalevskaya: Russian (1850–1891)

In every year since 1982, more women than men have earned bachelor's degrees in the United States. In every year since 2009, women have also earned a majority of doctoral degrees. But women did not always have equal opportunities when it came to higher education. The preference to males over females in education has been a marked feature since ancient societies. It wasn't until the mid to late nineteenth century that women's access to universities became widespread in the United States, largely as a result of the pressure produced by movements for women's rights. While men are now earning a minority of college degrees at all college levels, there is still male privilege to be found in academia; women are more likely to be found in lower-ranking academic positions. In 2015, women represented approximately half of assistant professors and associate professors in the United States but accounted for only a third of the full professors' ranks. In mathematics, the gap is even bigger: In 2015, women held only 15 percent of tenure-track positions in mathematics. Sofia Kovalevskaya (1850–1891) was a pioneer for women in mathematics, at a time when mathematics was an almost exclusively male-dominated field around the world, and it was widely believed that women had a natural disability for this subject. Furthermore, it was believed that if a woman undertook rigorous "brain work" such as mathematics, energy could be diverted from her reproductive system, threatening fertility and general well-being. Kovalevskaya was the first woman to obtain a doctorate (in the modern sense) in mathematics

Figure 38.1. Sofia Kovalevskaya.

and also the first woman appointed to a full professorship in modern Europe. Her name is still well-known among mathematicians for the Cauchy-Kovalevskaya theorem, a central result in the theory of differential equations.

Sofia Vasilyevna Kovalevskaya was born in Moscow on January 15, 1850, the second of three children. Both of her parents were well-educated members of the minor Russian nobility. Her father, Lieutenant General Vasily Vasilyevich Korvin-Krukovsky, served in the Imperial Russian Army as head of the Moscow Artillery. Her mother, Yelizaveta Fedorovna Schubert, descended from a family of German immigrants with a strong academic background. As it was common in her family's class at that time, Sofia was nursed by nannies and saw her parents only at dinner. During childhood, she did not have much contact with her siblings either, mainly because of the age difference: Her sister, Anna, was six years older than she, and her brother, Fjodor, was five years younger. Sofia was educated by governesses and private tutors, including native speakers of English, French, and German. When Sofia was eight years old, her father retired from the army and the family moved to Palibino, her father's family estate in the Vitebsk province. In the restoration of the estate, there was not enough wallpaper for Sofia's nursery and so the walls were papered with pages found in the attic. These were in fact notes of the Ukrainian mathematician Ostrogradski's (1801–1862) lectures on differential and integral analysis, left over from her father's student days. Sofia was curious about the mathematical notions and formulas on the wall in her room and they came to life when she overheard her uncle, an autodidact who read a lot of mathematics books, mention some of the terms she had seen on the wall. Sofia later wrote the following in her autobiography:

"The meaning of these concepts I naturally could not yet grasp, but they acted on my imagination, instilling in me a reverence for mathematics as an exalted and mysterious science, which opens up to its initiates a new world of wonders, inaccessible to ordinary mortals."

Her uncle fed her interest in mathematics and took the time to discuss the mathematical topics he was reading about with her. While Sofia was immediately fascinated with the concepts and ideas used in calculus, she was at first at little bored by her lessons in elementary geometry and algebra, which were provided by a private tutor. However, her attraction to mathematics in general began to grow as they moved on to more advanced material. In fact, it grew so strongly that her father decided to stop her mathematics lessons, but she continued to study mathematics on her own. At the age of fifteen, she read a physics book written by her neighbor Professor Tyrtov. When he visited the family, he realized that she had correctly interpreted some of the trigonometric formulas in the chapter on optics, without having been tutored in trigonometry. She had developed completely on her own some explanations of concepts such as the trigonometric sine function. Recognizing her mathematical talent, Tyrtov took quite some effort in trying to convince her father to let her study more advanced mathematics. He finally succeeded, and Sofia received private tutoring in calculus in St. Petersburg, where her family stayed most of the winter of 1866–1867. There she also met the Russian novelist Fyodor Dostoevsky (1821–1881), whom she admired. At that time, women in Russia, as well as in many other countries, were not allowed to attend lectures at the university, not even as guests. To continue her studies, Sofia would have to go abroad. However, traveling was also not an easy task, because women had no passports and needed the written permission of their father or husband to cross the border. Sofia entered into a marriage of convenience with Vladimir Kovalevsky, a young paleontology student, book publisher, and political radical, who was the first to translate and publish into Russian the works of Charles Darwin. The couple remained in St. Petersburg for only a few months and then went to Heidelberg, after a short stay in Vienna. Women could not matriculate at the University of Heidelberg, but Kovalevskaya persuaded the authorities to let her study there. In 1869, she was admitted as the university's first female student, although not with an official status, and with the condition that she would have to obtain the permission from each of her lecturers separately. She studied mathematics with Leo Koenigsberger (1837–1921) and, at his suggestion, moved to Berlin to continue her studies with Karl Weierstrass (1815–1897), one of the most famous mathematicians at that time. In spite of letters of recommendation from her professors in Heidelberg, Weierstrass wanted to assess her mathematical abilities personally. He gave her a difficult problem to solve, and when she presented her solution one week later, he was so impressed that he would not only accept her as his student, but also try to support her work. However, his advocacy was not enough for the university administration to allow her to attend his lectures. Over the next three years, therefore, Weierstrass taught her the content of his lectures privately, which actually turned the university's denial to her advantage; Kovalevskaya later wrote, "These studies had the deepest possible influence on my entire career in mathematics. They determined finally and irrevocably the direction I was to follow in my later scientific work: all my work has been done precisely in the spirit of Weierstrass."

By 1874, Kovalevskaya had completed three papers—on partial differential equations, on the dynamics of Saturn's rings, and on elliptic integrals—each of which was considered worthy of a doctorate by Weierstrass; with his support, she was granted her doctorate, summa cum laude, from Göttingen University. Kovalevskaya thereby became the first woman to have been awarded a doctorate at a European university. The first of the three papers was published in *Crelle's Journal* (one of the leading mathematics journals) in 1875. It contains what is now commonly known as the Cauchy–Kovalevskaya theorem, which proves the existence of local solutions of analytic partial differential equations under suitably defined initial conditions. The French mathematician Augustin-Louis Cauchy (1789–1857) had proved a special case in 1842, but the full result is due to Sofia Kovalevskaya. Because the meaning of this theorem cannot be accurately explained without assuming familiarity with advanced calculus, we will instead try to convey at least a vague idea of what it says by discussing an example. To understand the significance of this theorem, it is important to know that partial differential equations can be used to describe a wide variety of physical phenomena, such as sound, heat, electromagnetic waves, the motion of fluids, elasticity, and even quantum mechanics, as well as the curvature of space-time, including gravitational waves. Each of these phenomena is governed by a fundamental law of physics that can be represented mathematically by a partial differential equation. For example, if you pluck a guitar string, the string will vibrate and produce sound. The motion of the string is governed by a partial differential equation whose unknown is the elongation of the string from its equilibrium position, as a function of position and time. Depending on where you pluck the string and how much force you apply, the sound produced will have different pitch and volume. Imagine a snapshot of the deformed string just before you let it go—it will have the shape of a V as shown in figure 38.2 on the left. The exact shape, determined by the position of the vertex along the string and its distance from the equilibrium position, represents our initial condition of the partial differential equation for the vibrating string. If you now let the string go, it will start to oscillate, thereby taking on a sinusoidal shape as shown in figure 38.2 on the right.

This oscillatory motion is described by the solution of the partial differential equation. The Cauchy-Kovalevskaya theorem states that for suitable initial conditions, partial differential equations do have a solution and the solution is unique. In our example, this means that if we are able to solve the equation for the vibrating string for a given initial condition at time $t = 0$, we can predict exactly how the string will move—that is, how a snapshot of the string at any later time $t > 0$ will appear. But the Cauchy-Kovalevskaya theorem does not only apply to the vibrating string equation; it applies to a wide class of partial differential equations, in particular to many of those used in physics, and is therefore a very fundamental result. It does not tell

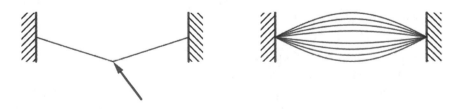

Figure 38.2.

us, though, how to obtain the solution, but tells us under which conditions a unique solution exists. This is important, because if we know that there is a unique solution, we can use it in an abstract sense to investigate its properties. In fact, for many partial differential equations it is not possible to calculate the solution explicitly, but one can find out a lot about its properties by assuming that it exists and then drawing conclusions from the equation it satisfies (for example, that the vibration of the string will decrease with time). This procedure of drawing conclusions about an object that we cannot calculate is justified by the Cauchy-Kovalevskaya theorem.

After her doctorate, Kovalevskaya went back to Russia, where she wanted to teach mathematics. However, women were not admitted to the required teacher certification examination, so the best job she was offered was teaching arithmetic in girls' elementary schools. With a bit of frustration, she completely turned away from mathematics, and she and her husband tried to become a conventional married couple. In 1878, their daughter, Sofia (called "Fufa"), was born. After almost two years that were devoted to raising her daughter, Kovalevskaya decided to resume her work in mathematics. Her husband, Vladimir, never got an academic position because of his radical beliefs, and their attempts to support themselves with real estate development failed, leading them to severe financial problems. Since Sofia still couldn't find an appropriate teaching position, she would now focus her energy on research. She started her endeavor by translating her six-year-old doctoral dissertation (which was written in German) into Russian and in 1880 presented the results at a scientific conference in Russia. In the same year, she moved with her husband and daughter to Moscow, where she visited seminars of the Moscow Mathematical Society. Her fascination for mathematics continued to grow ever stronger, whereupon, in 1881, she left her husband and went with her daughter to Berlin to continue with her research. She immersed herself in mathematical work and sent her daughter with a governess back to Russia, to her good friend Julija Lermontowa. In the meantime, Vladimir got involved with an oil company, which ruined him financially. He had always suffered severe mood swings, and in 1883, he committed suicide.

At that time, it was almost impossible for women to obtain a research position at a university. However, thanks to the Swedish mathematician Gösta Mittag-Leffler (1846–1927), who had known Kovalevskaya as a fellow student of Weierstrass, Kovalevskaya obtained a position as privat-docent at Stockholm University. In 1884, she was appointed to a five-year position as assistant professor and became an editor of *Acta Mathematica*, a mathematics journal established in 1882 by Gösta Mittag-Leffler, and now one of the most prestigious mathematics journals. In 1888, she won the Prix Bordin of the French Academy of Science for a work that contained the discovery of what is now known as the "Kovalevskaya top": The differential equation describing the motion of a spinning top cannot be solved explicitly for a general top, but there are three special cases for which an explicit solution can be calculated; the first of these three exceptions was discovered by Leonhard Euler, the second by Joseph-Louis Lagrange, and the third by Kovalevskaya.

She received further honors for her academic achievements in the following years and in 1889, she became full professor at Stockholm University. Kovalevskaya was the first woman in modern Europe to hold such a position. In the same year, she fell in love with Maxim Kovalevsky, who was distantly related to her deceased husband. However, she refused to marry him, because she knew that she would not be able to settle down and live with him. Indeed, mathematics was her first love, and it would also be her last. After the couple returned from a vacation in Nice, Kovalevskaya caught pneumonia and in 1891, at age forty-one, died of influenza at the peak of her mathematical powers and reputation. Kovalevskaya is not only remembered for her significant contributions to mathematics; she was also a talented writer. Her non-mathematical publications include a memoir, *A Russian Childhood* (1890) and a partly autobiographical novel, *Nihilist Girl* (1890).

CHAPTER 39

~

Giuseppe Peano: Italian (1858–1932)

If you have ever wondered where the symbols used in set theory came from, such as a symbol for union ($\cup$) and for intersection ($\cap$), then you need look no further than the famous Italian mathematician Giuseppe Peano, who was a prolific writer, and by many considered one of the founders of mathematical logic and set theory. His work involving the understanding of the characteristics of our natural numbers (1, 2, 3, 4, . . .) has remained with us through his Peano axioms, which we will consider after we take a quick look at his life story.

Giuseppe Peano was born on a farm in Spinetta, Piedmont, Italy, on August 27, 1858, where his parents, Bartolomeo Peano and Rosa Cavallo, worked the farm and which provided a three-mile walk for Giuseppe to his school. His uncle noticed that young Giuseppe was a very talented child and took him to Turin to begin his secondary schooling in 1870. By 1876, he enrolled at the University of Turin and graduated at the top of his class in 1880. He stayed at the university after graduation and eventually got to a position where he was the faculty member assigned to teaching calculus, apparently a significant honor at that time. His first written work was a textbook on calculus, which he published in 1848, which was then followed by a book on mathematical logic that has made him famous to this day; it was in this book that he introduced the modern symbols we use in studying sets, such as symbols for intersection and union, as mentioned above.

In 1887, Peano married Carola Crosio, during a time when he was also teaching at the Royal Military Academy, where he was later promoted to

Figure 39.1. Giuseppe Peano.

Figure 39.2. Giuseppe Peano and his wife Carola Crosio in 1887.

professor first-class. That was the time, in 1889, when he published his famous Peano axioms, which allow us to prove many relationships involving the natural numbers. As we mentioned earlier, Peano was a very prolific writer. The journal *Rivista di Mathematica*, which he founded, had its first publication in 1891. That same year he started a "Formulario Project," which was a compilation of all the known theorems and formulas used in mathematics; however, in this book he introduced his own notation. This led to some complications in the printing process, since he wanted all formulas to be printed on one line. As a result, he purchased his own printing press to ensure that this requirement was held firm.

In Paris, at the Second International Congress of Mathematicians in 1900, he met the famous British mathematician and logician Bertrand Russell (1872–1970), who was so impressed with his Formulario Project and the innovative logical symbols used therein that he left the conference earlier than planned just so he could read the book sooner. Russell then used Peano's logic notation in his later writings.

In 1901, when Peano was at the peak of his career, presenting at conferences and teaching calculus, differential equations, and vector analysis, he was also overly involved with his Formulario Project, so much so that his teaching began to weaken, and he was dismissed from the Royal Military Academy. However, he did retain his position at the University of Turin.

It is not uncommon for brilliant people to do things that are sometimes extraordinary or unusual. In 1903, Peano began to write in a form of Latin that he called *Latino sine Flexione*, which was later referred to as *Interlingua* and was based on a synthesis of Latin, German, English, and French vocabularies—however, with a very simplified type of grammar, removing all irregular forms. He did give speeches in this form of Latin and it was seen as a new language, one that served an international purpose.

Continuous work on the Formulario Project led to his publishing the fifth edition titled *Formulario Mathematico* in 1908. By this time the collection contained 4,200 theorems and formulas, along with justifications and proofs.

By 1910, Peano began to concentrate his efforts on writing mathematics texts for the secondary schools as well as a dictionary of mathematics. He also dabbled with international language issues. He continued to publish and to teach, eventually moving from infinitesimal calculus to complementary mathematics, which he felt better suited his style of mathematical thinking. He continued to teach at the University of Turin until he died of a heart attack in Turin on April 20, 1932.

The popular legacy that Peano has left for us are the *Peano axioms*, which are as follows:

1. There exists a natural number 1.
2. Every natural number has a unique successor, which is also a natural number.
3. The number 1 is not a successor of any natural number.
4. If two successors are equal, the numbers of which they are the successors are also equal.
5. If a set S of natural numbers contains 1, and if the successor of every number in S is also in S, then S is the set of all natural numbers.

It is the fifth axiom that renders us a form of proof called *mathematical induction*. This can then be restated thus: if a theorem is true for $n = 1$, and if the truth of the theorem for $n = k$ implies the truth of the theorem for $n = k + 1$, then the theorem is true for all positive integral values of n.

Let's apply this now to prove that the sum of all odd integers is equal to a square number. We would write this symbolically as

$S_n = \sum_{a=1}^{n} 2a - 1 = n^2$, or more simply written as:

$S_n = 1 + 3 + 5 + \cdots + (2n - 1) = n^2$.

Using Peano's fifth axiom, we must show that the relationship

$S_n = \sum_{a=1}^{n} 2a - 1 = n^2$ is true for $n = 1$. That is, $S_1 = 1^2 = 1$.

This time we will assume that the theorem is true for some value of n, such as k:

$S_k = 1 + 3 + 5 + 7 + \cdots + (2k - 1) = k^2$. We must now prove that if this theorem is true for $n = k$, it must also be true for the next consecutive value of n, which is $n = k + 1$. To do that we need to add the next odd integer $(2k + 1)$ to both sides of the above equation:

$S_k + (2k + 1) = 1 + 3 + 5 + 7 + \cdots + (2k - 1) + (2k + 1) = k^2 + (2k + 1)$.

We then get $S_{k+1} = k^2 + 2k + 1 = (k + 1)^2$.

Thus, we have proved that since the theorem was true for $n = 1$, and then we assumed it was true for $n = k$, and then showed it was true for the successor—namely, $n = k + 1$, we can conclude that it is true for all natural numbers. This is the very important legacy developed by Giuseppe Peano.

CHAPTER 40

David Hilbert:
German (1862–1943)

Until 1899, through most of the development of mathematics, it was felt that the work of Euclid was conclusive to our understanding of geometry. However, in 1899 the German mathematician David Hilbert published a book titled *Foundations of Geometry*, where he proposed a set of axioms that were to substitute for Euclid's traditional axioms. Among these axioms was the concept of *betweenness*, something with which Euclid did not concern himself. You might ask, how might this affect our study of geometry? To better understand how this concept firms up our study of geometry, we will begin by considering an example that could be easily shown in high school geometry and demonstrates—rather dramatically—what happens when we don't consider the concept of betweenness, as Euclid didn't.

Mistakes in geometry—also sometimes called fallacies—tend to come from faulty diagrams that result from a lack of definition. Yet, as we know, in ancient times some geometers discussed their geometric findings or relationships in the absence of a diagram. For example, in Euclid's work, the concept of "betweenness" was not considered. When not considering this concept, we can prove that *any* triangle is isosceles—that is, that a triangle that has three sides of different lengths actually *does* have two sides that are equal. This sounds a bit strange, but we can demonstrate this "proof," which yearns for the concept of betweenness.

We shall go through this short journey to expose this ridiculous result. So let's begin by drawing any scalene triangle (i.e., a triangle with no two sides of equal length) and then "prove" it is actually isosceles (i.e., a triangle with two sides of equal length). Consider the scalene triangle *ABC*, where we then draw the bisector of angle *C* and the perpendicular bisector of *AB*. From their point of intersection, *G*, draw perpendiculars to *AC* and *CB*, meeting them at points *D* and *F*, respectively. Depending on the shape of the scalene triangle drawn, we could now have four possibilities meeting the above description for the various scalene triangles: One possible configuration is shown in figure 40.1, where *CG* and *GE* meet inside the triangle at point *G*.

Another configuration is shown in figure 40.2, where *CG* and *GE* meet on side *AB*. (Points *E* and *G* coincide.)

A third configuration is shown in figure 40.3, where *CG* and *GE* meet outside the triangle (in *G*), but the perpendiculars *GD* and *GF* intersect the segments *AC* and *CB* (at points *D* and *F*, respectively).

Our fourth configuration is shown in figure 40.4, where *CG* and *GE* meet outside the triangle, but the perpendiculars *GD* and *GF* intersect the extensions of the sides *AC* and *CB* outside the triangle (in points *D* and *F* respectively).

The "proof" of the mistake or fallacy can be done with any of the above figures. Follow along and see if the mistake shows itself without

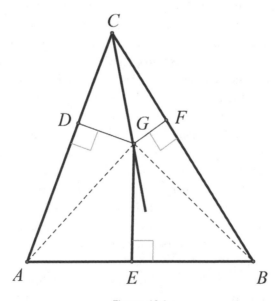

Figure 40.1.

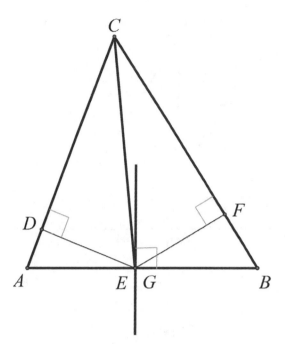

Figure 40.2.

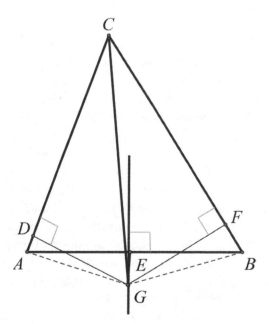

Figure 40.3.

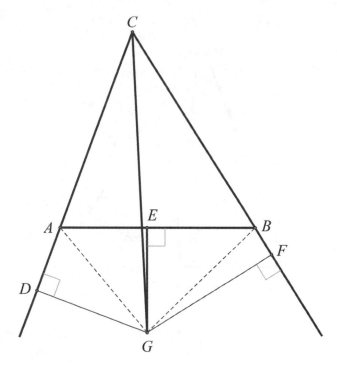

Figure 40.4.

reading further. We begin with a scalene triangle *ABC*. We will now "prove" that *AC* = *BC* (or that triangle *ABC* is isosceles).

As we have an angle bisector, we have ∠*ACG* ≅ ∠*BCG*. We also have two right angles, such that ∠*CDG* ≅ ∠*CFG*. This enables us to conclude that Δ*CDG* ≅ Δ*CFG* (SAA). Therefore, *DG* = *FG*, and *CD* = *CF*, since a point on the perpendicular bisector (*EG*) of a line segment is equidistant from the endpoints of the line segment *AG* = *BG*. Also ∠*ADG* and ∠*BFG* are right angles. We then have ∠*DAG* ≅ ∠*FBG* (because they have a respective hypotenuse and leg congruent). Therefore *DA* = *FB* . It then follows that *AC* = *BC* (by addition in figs. 40.1, 40.2, and 40.3; by subtraction in fig. 40.4).

At this point you may feel quite disturbed. You might challenge the correctness of the figures. Well, by rigorous construction you will find a subtle error in the figures. We will now divulge the mistake and see how it leads us to a better and more precise way of referring to geometric concepts, something Hilbert's axioms attempted to do.

First, we can show that the point *G must* be outside the triangle. Then, when perpendiculars meet the sides of the triangle, one will meet a side

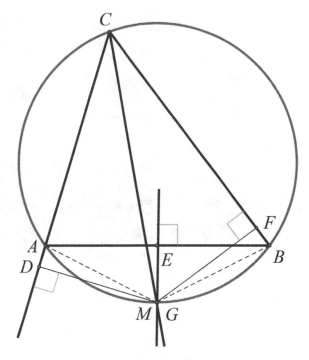

Figure 40.5.

between the vertices, while the other will not. We can "blame" this mistake on Euclid's neglect of the concept of betweenness.

Begin by considering the circumcircle of triangle *ABC* (fig. 40.5). The bisector of angle *ACB* must contain the midpoint, *M*, of arc *AB* (because angles *ACM* and *BCM* are congruent inscribed angles). The perpendicular bisector of *AB* must bisect arc *AB*, and therefore, pass through *M*. Thus, the bisector of angle *ACB* and the perpendicular bisector of *AB* intersect on the circumscribed circle, which is *outside* the triangle at *M* (or *G*). This eliminates the possibilities we used in figures 40.1 and 40.2.

Now consider the inscribed quadrilateral *ACBG*. Since the opposite angles of an inscribed (or cyclic) quadrilateral are supplementary, $\angle CAG + \angle CBG = 180°$. If angles *CAG* and *CBG* were right angles, then *CG* would be a diameter and triangle *ABC* would be isosceles. Therefore, since triangle *ABC* is scalene, angles *CAG* and *CBG* are not right angles. In this case one must be acute and the other obtuse. Suppose angle *CBG* is acute and angle *CAG* is obtuse. Then in triangle *CBG* the altitude on *CB* must be *inside* the triangle, while in obtuse triangle *CAG*, the altitude on *AC* must be *outside*

Figure 40.6. David Hilbert in 1910 and 1940.

the triangle. The fact that one and *only one* of the perpendiculars intersects a side of the triangle *between* the vertices destroys the fallacious "proof." This demonstration hinges on the definition of betweenness, a concept not available to Euclid, but made clear through Hilbert's axioms.

David Hilbert was born on January 23, 1862, in Königsberg, Prussia, which today is Kaliningrad, Russia, but then a German city. His father, Otto Hilbert, was a city judge and his mother, Maria Hilbert, pursued philosophy and astronomy. With this rearing, David Hilbert, as a child, already showed a special gift for mathematics and an interest in languages. In 1872, he entered the Friedrichs Kolleg Gymnasium, and seven years later graduated from the Wilhelm Gymnasium. The following year, in 1880, he enrolled in the University of Königsberg to study mathematics. There he befriended a colleague mathematician, Hermann Minkowski (1864–1909), who in 1882 returned from Berlin to Königsberg, where he had been previously studying. Minkowski became Hilbert's dearest lifelong friend. In 1884, Hilbert and Minkowski collaborated with a newly arrived professor from Göttingen, Adolf Hurwitz (1859–1919). This three-way friendship and collaboration had a lasting effect on their professional careers. Upon receiving his doctorate in 1885, Hilbert began to prepare for the state examination to qualify for a teaching position at a gymnasium. Of course, he passed the examination. During this time, he also attended courses on plane geometry and spherical geometry. Hurwitz suggested that he spend the winter of that year at the University of Leipzig, specifically to attend the lectures of the well-known German mathematician Felix Klein (1849–1925). Klein then suggested that he visit Paris, to establish contact with several famous mathematicians, which he did successfully, even though it was somewhat strenuous for his French colleagues to speak German, since Hilbert was unable to speak French. Soon thereafter, he returned to the University of Königsberg to be a member of the faculty from 1886 until 1895, being appointed to the full professorship in 1893.

On October 12, 1892, Hilbert married his second cousin, Käthe Jerosch. They had one son, Franz, who was born on August 11, 1893. Hilbert's professional career moved along with the strong support of Felix Klein, who arranged for Hilbert to be appointed to the chair of mathematics at the University of Göttingen, which was a center for many famous mathematicians such as Carl Friedrich Gauss (see chap. 29), Bernhard Riemann (see chap. 36), and Emmy Noether (see chap. 42), to name just a few. Hilbert spent the rest of his professional life at the University of Göttingen, where he supervised 69 doctoral students, many of whom became famous mathematicians in their own right. Hilbert also showed an intense interest in mathematical physics, which built a strong physics component at the university, and which ultimately showed its significance in having generated three Nobel laureates in physics: Max von Laue (1914), James Franck (1925), and Werner Heisenberg (1932).

In 1899, Hilbert published a book titled *Foundations of Geometry*, where he proposed a set of axioms that were intended to replace those that Euclid made famous in his *Elements*. Over the next several years, the book was translated into several languages. This set a new trend of a modern axiomatic method. One newly introduced concept was that of betweenness, as we mentioned earlier.

In the Appendix (see page 401), we provide a summary of Hilbert's Axioms, but note that line segments, angles, and triangles may each be defined in terms of points and straight lines, using the relations of betweenness and containment. All points, straight lines, and planes in the following axioms are distinct, unless otherwise stated. Hilbert's axioms essentially unified plane geometry and solid geometry into a single system.

In 1900, at the *Second International Congress of Mathematicians* in Paris,[1] Hilbert proposed his famous 23 unsolved problems, which were considered the most challenging problems ever produced by a mathematician.

Hilbert surveyed many fields of mathematics as he developed this list of 23 research problems that he thought would be a significant challenge to mathematicians going into the 20th century. By the way, one of his stated problems was Goldbach's Conjecture (see chap. 21). Unfortunately, the problems proposed are beyond the scope of this book.[2] However, suffice it to say, many of the problems were solved during the twentieth century.

Hilbert was also very concerned with logical reasoning, which certainly set the course of formalistic foundations of mathematics for the twentieth century and beyond. His work around 1910 with integral equations eventually led to research in functional analysis, which has had significant applications in physics. About this time, he also proved a conjecture in number theory—namely, that all positive integers can be expressed as the sum of a certain number of nth powers. For example, $11 = 3^2 + 1^2 + 1^2$, where $n = 2$, or as another example, $65 = 4^3 + 1^3$, where $n = 3$.

Throughout the early part of the twentieth century, Hilbert remained a major contributor in the field of mathematics, as evidenced that from 1902 until 1939, he was the editor of *Mathematische Annalen*, one of the world's leading mathematical journals. In 1925, at age 63, Hilbert contracted the vitamin deficiency disease pernicious anemia, which left him rather exhausted and he never produced as much after that time as he did in his previous years.

Another detrimental aspect of this brilliant mathematician's career occurred in 1933, when the Nazis removed many significant Jewish faculty members from the University of Göttingen. During the following year, Hilbert had occasion to meet the German minister of education, Bernhard Rust, who asked Hilbert if the Mathematical Institute had suffered much as a result of the removal of the Jewish faculty members. Hilbert replied, "Suffered? It doesn't exist any longer, does it!"[3] Initially, Hilbert spoke out vehemently against the Nazi repression of his Jewish mathematician friends. However, in time, when he saw that his disgust was futile, he remained silent and withdrawn. Although he was raised as a Calvinist, he later left the church and became an agnostic. Hilbert died on February 14, 1943, in relative obscurity; most of his colleagues at the University of Göttingen were gone.

There are many aspects of mathematics for which Hilbert is remembered today. One such is known as a Hilbert space, which is a generalization of a Euclidean space. This extends the methods of vector algebra and calculus to spaces with any finite or infinite number of dimensions. Hilbert spaces provided the basis for important contributions to physics over the following decades and may still offer one of the best mathematical formulations of quantum mechanics. We show an example of a Hilbert algorithm for space-filling curves in figure 40.7.[4]

The mid-1930s was also the time when Hilbert and the Swiss mathematician Paul Bernays (1888–1977) coauthored the two-volume work *Foundations of Mathematics*, which presented fundamental mathematical ideas and introduced a collection of axiomatic systems, which formalized natural numbers and their subsets, and which offered an alternative to axiomatic set theory.

Hilbert's legacy lives on today because of the many innovations he made in a variety of fields of mathematics. This is evidenced by the many mathematical concepts that still carry his name, such as Hilbert Number, Hilbert Matrix, Einstein–Hilbert Equations, Hilbert's Axioms, Hilbert System, Hilbert Polynomial, Hilbert Function, Hilbert Curves, and many others.

Hilbert produced algorithms for space-filling curves which completely fill up higher dimensional spaces, such as squares or cubes.

A recursive algorithm can be used repeatedly to create a continuous and non-intersecting curve such that every pixel in a grid is traversed once, and only once. The illustrations below show the first six iterations in such a process.

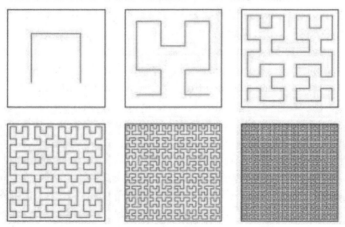

Figure 40.7.

~

CHAPTER 41

G. H. Hardy:
English (1877–1947)

In today's technical world, mathematics is seen as the key to advancement, which means that mathematics is being received and appreciated for its usefulness not only in the sciences and in computer science, but also in finance and other logic-based fields. As a result, the curricula in many schools today, which should show mathematics through its power and beauty, focuses largely on the former and not so much on the latter. England's most famous mathematician of the twentieth century, G. H. Hardy spent his life championing the notion that mathematics should be appreciated for its beauty, and not necessarily for its usefulness, although some of his research has applied mathematics to solve problems in fields such as genetics. As a matter of fact, in 1940, he wrote an essay on the aesthetics of mathematics, which is still available today as a book titled *A Mathematician's Apology*, where he tries to give the layman insights into the mind of a mathematician. The overarching theme is that mathematics should be appreciated for its own beauty, rather than for its usefulness to solve problems in other fields. Thus, he enthusiastically pursued studying pure mathematics as opposed to applied mathematics, where the latter was an abhorrence to him, especially when it was applied to military strategies and maneuvers.

Godfrey Harold Hardy was born on February 7, 1877, in Cranleigh, Surrey, England, to parents who were both educators and with a special talent in mathematics. Unfortunately, because of a lack of funding, they were not able to afford a university education. Perhaps somewhat motivated by

Figure 41.1. G. H. Hardy.

his parents, Hardy showed early signs of mathematical talent when at the age of two he was able to write the sequence of numbers from 1 to 1,000,000. He went to school in his hometown up to age 12, and in 1889 won a scholarship to Winchester College, Winchester, England. This college was considered at the time to offer the best training in mathematics in England; however, Hardy found nothing enjoyable there beyond the academic training. Hardy was relatively frail and shy as compared to his colleagues and found beating them in mathematics gave him some posture. In 1896, Hardy entered Trinity College, Cambridge University on a scholarship. At the start he was assigned to Robert Rumsey Webb (1850–1936) as his coach, who seemed more interested in showing him how to pass examinations than to make the subject of mathematics interesting and exciting. As he was contemplating a change of subject interest to history, he had the good fortune of having a new coach, A. E. H. Love, who guided him to read material that once again rekindled his interest in mathematics. In later years, Hardy claims that Camille Jordan's (1838–1922) book, *Cours d'analyse*, had a lasting effect on him and defined for him what mathematics really meant.

In 1898, Hardy graduated as the fourth-highest level mathematics student from Cambridge University's graduating class, which disappointed him greatly, because he thought he should be at the top of the graduating class. Yet in 1900, he was elected a fellow of Trinity College, and then the following year was awarded Smith's Prize for Excellence in Mathematics. Over the next decade, he wrote many professional papers on the convergence of series and integrals and other allied topics. However, his most distinctive work during this period was in 1908 where he published *A Course of Pure Mathematics*, which was the first rigorous English exposition of number, functions, limits, etc. This book was aimed at the undergraduate population and was a major transformation of the curriculum to that point. Hardy was very modest about his work and in retrospect in later years, he was not too enchanted with his mathematics production during that time.

In 1911, Hardy began to collaborate with John E. Littlewood (1885–1977), which was to be a relationship that lasted 35 years. They worked on mathematical analysis and analytical number theory. They were involved in grappling with Waring's problem, which states that a natural number n has an associated positive integer k, such that every natural number is the

Figure 41.2. Hardy and Littlewood in 1924.

sum of the at most k natural numbers to the power of n. For example, every natural number is the sum of at most 4 squares, 9 cubes, or 19 fourth powers. The problem was posed by the British mathematician Edward Waring (1736–1798) in 1770 and was proved to be true by David Hilbert in 1909. It served as the basis for further investigations by Hardy and Littlewood, who further made a number of conjectures.

One of their conjectures dealt with twin prime numbers, which are prime-number pairs that are consecutive odd prime numbers, for example, 5 and 7 are twin primes, as are 41 and 43. They took this a step further to investigate sequences of primes with a common difference between them. Another of their conjectures concerns the number of primes in intervals. The conjecture states that $\pi(x+y) \le \pi(x) + \pi(y)$, where $\pi(x)$ represents the number of prime numbers less than or equal to the real number x. For example, $\pi(2) = 1$, since there is only one prime number less than or equal to 2—namely, 2 itself. Thus, the conjecture implies that $\pi(y+2) \le 1 + \pi(y)$, or $\pi(y+2) - \pi(y) \le 1$, meaning that the number of primes greater than y and less than or equal to $y + 2$ is at most 1. This is correct because at most one of two consecutive numbers can be prime. However, the conjecture even states that $\pi(x+y) - \pi(x) \le \pi(y)$, which means that the number of primes greater than $\pi(x)$ and less than or equal to $\pi(x+y)$ is not larger than the number of primes between 1 and y. In other words, the number of primes among n consecutive numbers gets smaller as the starting number gets larger. Over the next several decades this collaboration between Hardy and Littlewood was considered one of the major accomplishments in the field of mathematics. Today there are seven volumes published by Oxford University Press consisting of Hardy's collected papers,[1] many of which are collaborations with Littlewood and Ramanujan, as well as with other famous mathematicians.

By 1913, life began to change for Hardy. He received a letter from the Indian mathematics enthusiast Srinivasa Ramanujan asking to obtain support with his studies. Previously, Ramanujan's efforts to obtain support were ignored by two other eminent mathematicians, but somehow Hardy recognized the genius from the received correspondence and invited Ramanujan to visit him at Cambridge, England. This led to an important collaboration that resulted in five very significant mathematical papers. (See chap. 43 for more about this collaboration.) One example of Hardy's collaboration with Ramanujan is known as the Hardy-Ramanujan asymptotic formula, which has broad applications in physics. This work was based on integer

Figure 41.3. Ramanujan (center) and Hardy (right).

partitions, which are ways of writing positive integers as a sum of other integers. That is, two sums that differ only by which numbers are added to arrive at the sums are considered the two partitions of the same sum. For example, the number 4 can be partitioned in five distinctive ways: 4, 3 +1, 2 + 2, 2 + 1 + 1, 1 + 1 + 1 + 1.

This was roughly the time when his research partner Littlewood left Cambridge to join the Royal artillery in World War I. Hardy would have followed him, but he was rejected from military service on medical grounds. However, by 1919, relatively unhappy at Cambridge University, he took the position of Savilian Professor of Geometry at Oxford University. This began a productive stretch of time; he was once again collaborating with Littlewood, who remained at Cambridge University.

Although Hardy prided himself on concentrating on pure mathematics, he also got involved in solving problems a bit outside of the mathematical realm. One such was an issue raised by the German obstetrician Wilhelm Weinberg in 1908, which Hardy perfected to form the Hardy-Weinberg principle, and which states that allele and genotype frequencies in a population will remain constant from generation to generation in the absence of other evolutionary influences.

Although Hardy never married, he did have some relationships, yet he was a rather shy person who tried not to be the center of attention, even when he was being honored. Despite his shyness he had developed friendships with well-known mathematicians and philosophers of his day, such as Bertrand Russell, John Maynard Keynes, G. E. Moore, George Polya, and others. He was a member of several societies in his later years, such as the Cambridge Apostles and the Bloomsbury Group.

He felt that his most productive years were his young years. In his famous paper *A Mathematician's Apology*, he cited other famous mathematicians whose most productive years were before age 40, such as Galois, who died at age 20; Abel, who died at 27; Ramanujan, who died at age 33; and Riemann, who died at 40. Yet he did mention exceptions such as Gauss, who published a work on differential geometry at age 50, but he claims to have done the work at age 40. (Perhaps that could account for the rule that in order to qualify for the most prestigious prize in mathematics, the Fields Medal, one cannot be older than age 40.) We should take note that the conqueror of Fermat's Last Theorem, the British mathematician Andrew Wiles, performed this feat at age 42.

At the end of World War II in 1945, Hardy's health began to weaken as did his creativity, and as a result he became rather depressed. Even walking became a chore for him, and he was forced to use other forms of transportation such as taxis. In mid-1947 he tried to commit suicide by taking an overdose of barbiturates. This did not take his life but made him rather ill and bedridden. He told his friends that he simply wanted to die, but he did not choose to attempt suicide again since he was not good at it. His sister looked after him in his last days and he died on December 1, 1947, in Cambridge, England.

A few weeks before he died, Hardy received the Copley Medal of the Royal Society "for his distinguished part in the development of mathematical analysis in England during the last 30 years." It should be noted that he was seen in England as a leading figure in mathematics, as evidenced by the fact that he was president of the London Mathematical Society from 1926 to 1928 and again from 1939 to 1941. And in 1929 the society awarded him the De Morgan Medal, which was a great honor.

Throughout his life, Hardy was an atheist and loved the sport of cricket. In fact, the two loves of his life were mathematics and cricket. We mentioned his shyness, which also manifested itself in his not allowing his photograph to be taken, so there are believed to be only five photographs ever taken of

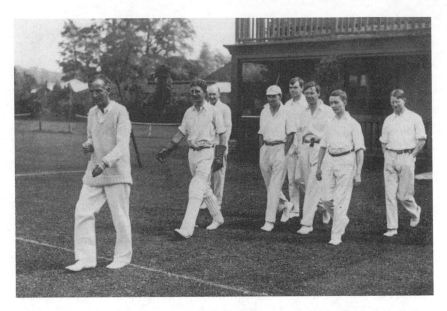

Figure 41.4. Hardy leading his cricket team onto the field.

him. He also despised having a mirror in his midst, and it is said that whenever he entered a hotel room with a mirror, he immediately covered it with a towel. He was very devoted to his students and expected perfection from them, yet he felt that one of his greatest contributions to mathematics was discovering Ramanujan.

Perhaps in closing the biography of Godfrey Harold Hardy it would be appropriate to consider a quotation from his essay *A Mathematician's Apology*:

> I emphasised the permanence of mathematical achievement—What we do may be small, but it has a certain character of permanence; and to have produced anything of the slightest permanent interest, whether it be a copy of verses or a geometrical theorem, is to have done something utterly beyond the powers of the vast majority of men. And—In these days of conflict between ancient and modern studies, there must surely be something to be said for a study which did not begin with Pythagoras, and will not end with Einstein, but is the oldest and the youngest of all.

CHAPTER 42

~

Emmy Noether:
German (1882–1935)

You have probably seen an ice skater spinning on the tip of one skate, and suddenly start spinning dramatically faster as she pulls her limbs closer to her body. This faster rotation results from a redistribution of mass. You can make yourself suddenly spin faster while sitting in a rotating desk chair. Sit in the chair and hold your arms and legs straight out, and have a friend give you a gentle spin. While you are spinning slowly, quickly pull the masses in toward your body and notice that you rotate much faster. If you stick out your arms and legs, you will slow down again. The spinning desk chair is a demonstration of the "conservation of angular momentum," which is one of the fundamental conservation laws in physics. It is similar to the conservation of linear momentum, which is more familiar to most people. Newton's first law states that every object will remain at rest or in uniform motion in a straight line unless it is acted upon by a force. Today, we call this observation the law of conservation of momentum. The linear momentum, p, of an object with mass m and velocity v is the product mv. It is a conserved quantity, meaning that its value and direction remain constant as long as no force is applied. Similarly, a rotating object tends to remain rotating with a constant angular momentum unless it is acted upon by an outside twisting force. If an object with mass m rotates with an angular velocity w, then its angular momentum, L, is the product mwr^2, where r is the radius of the circle that the object traces out. In a closed physical system, the total angular momentum is conserved, meaning that its value and the axis of rotation remain constant. If you are sitting in a spinning desk chair with your arms and legs stretched out, these parts of your body will trace out circles of certain radii as the chair rotates. If you now pull your arms and legs closer to your body, the radii of these circles get smaller. But since the total angular momentum of the spinning chair is conserved—meaning that the value of the product mwr^2 remains constant—a smaller radius implies that the angular velocity increases. For example, if the radius is halved, w will quadruple. Thus, the increase in angular velocity as you pull your arms and legs to your body is a consequence of the conservation of angular momentum. Note, however, that we have not taken friction into account here. Friction acts like a force counteracting the rotation; it will gradually make the chair spin more slowly until it eventually stops. Linear and angular momentum are not the only quantities in physics satisfying conservation laws; other examples include energy and mass. Conservation laws have always been very important in physics, yet they often appeared somewhat miraculously

Figure 42.1. Emmy Noether.

as not-at-all-obvious consequences of the governing equations of a physical theory. However, in the early twentieth century the German mathematician Emmy Noether made a discovery that provided both deeper insight into conservation laws and a practical calculation tool to find all conservation laws within a given physical theory. Noether proved that if a physical system has a symmetry property, then there is always an associated conservation law. For example, a system is symmetric under rotations if it behaves the same, regardless of how it is oriented in space. Noether's theorem then tells us that the angular momentum of the system is conserved. Noether's theorem was a milestone in theoretical physics; her work continues to be relevant to the development of theoretical physics and mathematics. Albert Einstein described her as the most important woman in the history of mathematics. Let us now find out who this spectacular woman was.

Amalie Emmy Noether was born on March 23, 1882, in the town of Erlangen, Germany. Her first name was "Amalie," after her mother Ida Amalia Kaufman, but already at a very young age she started using her middle name, "Emmy." Both her parents came from wealthy Jewish merchant families. Emmy was the first of their four children and the only girl. Her father, Max Noether, was a renowned mathematician and a professor at the University of Erlangen. Yet Emmy did not show any particular interest in mathematics in elementary school. Furthermore, she was nearsighted and had a slight lisp, and so did not stand out in any way. At high school she studied German, English, French, and arithmetic. She was also taught to cook and clean and took piano lessons. High schools were not coeducational at that time, and the curriculum of girls' schools was tailored to their future role as housewives. Emmy loved to dance, but otherwise she was not very enthusiastic about activities that were considered typical for girls. Mathematics was not intensively taught at her school. She finished high school in 1897, and after further study of English and French she took the Bavarian State teacher certification examinations; in 1900, she became a certificated teacher of English and French in Bavarian girls' schools. Although she was now qualified to teach languages at girls' schools, she decided not to pursue this path any further, but instead to aim at a higher education. Against many obstacles, Noether continued her studies at the University of Erlangen, where she was one of only two female students. Women were not officially allowed to study at universities, and therefore she had to get permission from each professor whose lecture she wanted to attend. Her focus had meanwhile shifted to mathematics, although she was still interested in studying languages as well. In 1903, she passed the entrance examination that would have allowed a male student to enter any university. She went to the University of Göttingen, which was the leading place for mathematical research in Germany at that time. While in Göttingen, she attended lectures by Karl Schwarzschild, Otto Blumenthal, David Hilbert, Felix Klein, and Hermann Minkowski. Again, she was only allowed to audit lectures as a guest, without being officially enrolled. After one semester at Göttingen, in 1904, she returned to Erlangen, when the university finally allowed women to enroll. She declared her intention to study mathematics and was then among the first women in Germany to officially study at a university. She completed her dissertation under the supervision of Paul Gordan in 1907. The natural next step would have been the habilitation, a postdoctoral qualification that was required for a professorship. Of course, this track was not possible for Emmy Noether; restrictions on women's access to universities at these higher levels were still in effect. For the next seven years she taught at the University of Erlangen's Mathematical Institute without pay, assisting her father and substituting for him when he did not have time to hold a scheduled lecture. She also continued her research and published papers extending the work of her thesis. The quality of her work made her name known to other mathematicians, and in 1909, Noether became a member of the German Mathematical Society and was invited to give a lecture at its annual meeting. Naturally, she would stand out at such extremely male-dominated events. During a mathematical conference in Vienna in 1913, she visited the Austrian mathematician Franz Mertens (1840–1927), whose grandson later remembered her visit, describing her as follows:

> Although a woman, [she] seemed to me like a Catholic chaplain from a rural parish—dressed in a black, almost ankle-length and rather nondescript, coat, a man's hat on her short hair . . . and with a shoulder bag carried crosswise like those of the railway conductors of the imperial period, she was rather an odd figure.

In 1915, Noether was invited to Göttingen by David Hilbert and Felix Klein, who needed her help in understanding certain aspects of general relativity, a geometrical theory of gravitation developed by Albert Einstein (1879–1955). They encouraged her to apply for a habilitation at the University of Göttingen. However, she received strong protests from numerous other faculty members. Reportedly, one of the professors asked, "What will our soldiers think when they return to the university and find that they are required to learn at the feet of a woman?"—to which Hilbert gave a now-famous reply: "We are a university, not a bath house!" Yet Hilbert's efforts turned out to have been in vain, when the responsible authorities rejected a petition of the faculty to grant her the habilitation to become a privat-docent. Without an official position, she could not receive any payment from the university and her lectures were advertised under Hilbert's name, with Noether as his "assistant." Without her family's financial support, she would not have been able to continue her research at the University of Göttingen. Soon after her arrival at Göttingen, she proved the theorem now known as Noether's theorem, which was, however, not published until 1918. Upon receiving her work, Einstein wrote to Hilbert,

> Yesterday I received from Miss Noether a very interesting paper on invariants [conserved quantities]. I'm impressed that such things can be understood in such a general way. The old guard at Göttingen should take some lessons from Miss Noether! She seems to know her stuff.

In 1918, after the end of World War I and the collapse of the German empire, Germany became a republic and women's rights were significantly improved, including the admission of women to the habilitation process. In 1919, Emmy Noether became the first woman to be granted a habilitation, allowing her to obtain the rank of privat-docent—however, still without a salary. It was not until 1922 that she became what translates to an "associate professor without tenure," and began to receive a modest salary. Until then she had to live off a small inheritance, adopting a frugal lifestyle that she would maintain for the rest of her life. In 1924, a young Dutch mathematician, B. L. van der Waerden (1903–1996), began working with Noether, who provided fundamental ideas of abstract conceptualization. In 1931, he published *Moderne Algebra*, an influential two-volume treatise on abstract algebra. The second volume is based heavily on Noether's work. Although Noether did not seek recognition, van der Waerden included the following as a note in the seventh edition: "based, in part, on lectures by E. Artin and E. Noether." Noether remained a leading member of the University of Göttingen mathematics department until 1933, during which time she had visiting professorships in Moscow and Frankfurt-am-Main, Germany. However, despite her major contributions in the field of mathematics, she was never promoted to full professor at the University of Göttingen. In 1933, Hitler became the German Reichskanzler and the Nazi administration immediately began to remove Jews and politically suspect government employees from their jobs, including university professors. Moreover, anti-Semitic attitudes, from both students and colleagues, made the universities hostile environments for Jewish professors. Noether was dismissed by the Prussian Ministry for Sciences, together with several of her colleagues at the University of Göttingen, including Richard Courant, who later founded an institute for graduate studies in applied mathematics in New York City, now known as the famous Courant Institute of Mathematical Sciences at New York University. When the Nazi Party came to power, a large number of German professors lost their employment, among them several future Nobel laureates. Their colleagues in the United States and in other countries sought to provide job opportunities for them, mainly to the United States and Great Britain, leading to a historic emigration of brain power from continental Europe. In fact, the rise of fascism in Europe in the 1930s put an end to Europe's cultural and intellectual supremacy. The exiles from Germany and, later, Austria, Italy, and France, enabled the United States to gain supremacy in many sciences and mathematics. Noether was contacted by representatives of two educational institutions: Bryn Mawr College in Pennsylvania, and Somerville College at University of Oxford, England. She accepted a one-year visiting professorship at Bryn Mawr College, where she was made very welcome by Anna Johnson Pell Wheeler, who was head of mathematics and who had studied at the University of Göttingen in 1905. In 1934, Noether was invited to give weekly lectures at the Institute for Advanced Study in Princeton. However, she remarked about Princeton that she was not welcome at "the men's university." In April 1935, doctors discovered a tumor in Noether's pelvis. She was operated on two days later. Although the operation seemed to have been successful at first, she collapsed a few days later and died on April 14, 1935, in Bryn Mawr, Pennsylvania. Noether never married and had no children.

Emmy Noether is consistently ranked as one of the greatest mathematicians of the twentieth century. At an exhibition at the 1964 World's Fair in New York City, an exhibition devoted to Modern Mathematicians, Noether was the only woman represented among the notable mathematicians of the modern world. In a letter to the editor of the *New York Times*, published a few weeks after her death, Albert Einstein wrote:

In the judgment of the most competent living mathematicians, Fräulein Noether was the most significant creative mathematical genius thus far produced since the higher education of women began. In the realm of algebra, in which the most gifted mathematicians have been busy for centuries, she discovered methods which have proved of enormous importance in the development of the present-day younger generation of mathematicians.

Although Noether's theorem had a huge impact on the development of theoretical physics, among mathematicians she is perhaps best remembered for her contributions to abstract algebra. In his introduction to Noether's collected papers, Nathan Jacobson wrote that "The development of abstract algebra, which is one of the most distinctive innovations of 20th-century mathematics, is largely due to her—in published papers, in lectures, and in personal influence on her contemporaries."

CHAPTER 43

~

Srinivasa Ramanujan: Indian (1887–1920)

Although many books have been written about famous mathematicians, not many are so lauded as to have a full-length film produced about their lives. The title of the film *The Man Who Knew Infinity*[1] is the same title as the book by Robert Kanigel[2] on which this film is based. It highlights the short life of the genius mathematician Srinivasa Ramanujan, who could claim no formal higher education and yet showed a brilliance that allowed him to be accepted by the leading British mathematicians of his day; one such was G. H. Hardy. A cute anecdote that can illustrate the brilliance of this mathematician occurred shortly before he died and was lying ill in a hospital in London. His now close friend Hardy went to visit him in the hospital and recalled a story that describes his brilliance in simple terms[3]: "I remember once going to see him when he was ill at Putney. I had ridden in taxi cab number 1729 and remarked that the number seemed to me rather a dull one, and that I hoped it was not an unfavorable omen. 'No,' he replied, 'it is a very interesting number; it is the smallest number expressible as the sum of two cubes in two different ways.'" Through his unique brilliance he was able to immediately notice that $1729 = 1^3 + 12^3 = 9^3 + 10^3$. Quite astonishing!

Srinivasa Ramanujan was born on December 22, 1887, in what is today Tamil Nadu, India, at the home of his maternal grandparents. He grew up in the town of Kumbakonam, India, in a house that today is a museum to memorialize his achievements. He ended up being an only child, since his three siblings died within a year of their birth. Although at age two he did contract smallpox, he survived, unlike many others at that time who had the same disease. During his youth he lived with both maternal and paternal grandparents, who against his will sent him to school. The school board dismissed him and he went back to live with his parents and formed a close relationship with his mother. Once again in primary school, he performed well in English and other subjects, but excelled in arithmetic. From there he entered secondary school in Kumbakonam. There, he had his first opportunity to be exposed to mathematics beyond arithmetic. At age eleven he was already at college-level mathematics and by age thirteen he mastered advanced trigonometry, during which time he was already developing sophisticated mathematical theorems. Within the next year, he was already receiving awards for his mathematical achievements and showed a specific interest in geometry and infinite series. When Ramanujan was fifteen, he was shown how to solve a cubic equation, and he devised his own technique for solving quadratic equations. In 1903, when Ramanujan was sixteen years old, he got a copy of *A Synopsis of Elementary Results in Pure and Applied Mathematics* from the library, which is a

Figure 43.1. Srinivasa Ramanujan.

collection of 5,000 mathematical theorems by G. S. Carr. It is believed that this book inaugurated his genius for mathematics.[4]

Ramanujan's brilliance began to manifest itself when he further investigated the Bernoulli numbers; when he tried to explain it to his peers, he found that it was far beyond their ability to comprehend what he was saying. The Bernoulli numbers is a sequence of rational numbers often used in number theory. The first twenty Bernoulli numbers are

$$1, \pm\frac{1}{2}, \frac{1}{6}, 0, -\frac{1}{30}, 0, \frac{1}{42}, 0, -\frac{1}{30}, 0, \frac{5}{66}, 0, -\frac{691}{2730}, 0, \frac{7}{6}, 0, -\frac{3617}{510}, 0, \frac{43867}{798}, 0, -\frac{174611}{330}.$$

(See chap. 22.)

In 1904, he entered the Government Arts College in Kumbakonam, where he excelled in mathematics, but found absolutely no interest in any of the other subjects. As a result, by 1905 he dropped out of school and left home, eventually enrolling at Pachaiyappa's College in Madras, India. He was no more successful there than previously, failing all exams except mathematics, where he continued to excel. He eventually left this college without a degree so that he could pursue further research in mathematics. He lived in extreme poverty with the threat of starvation regularly hanging over him. It was not until 1910 that the founder of the Indian Mathematical Society, Professor Ramaswami, recognized his talents and brought him as a researcher to Madras University.

Life moves on; on July 14, 1909, Ramanujan married a girl (Janakiammal), who was selected by his mother and who was only ten years old at the time. This was an Indian custom and not unusual. His new wife stayed at home until she reached puberty and was then allowed to live with her husband. Now with family responsibilities, Ramanujan was in search of a job—in particular, a clerical position. In the meantime, he sustained himself by tutoring students at Presidency College. During this time, Ramanujan fell seriously ill a few times and each time was saved by physicians offering their services pro bono. During his last sickness, he thought he would not survive and gave his notebooks to colleagues for safekeeping, but when he recovered he retrieved his books and continued his research. In 1912, along with his mother and wife he moved to George Town, Madras. Finally, in May 1913 he was able to get a research position at Madras University and moved to Triplicane with his family. Through a variety of recommendations, and despite concerns of the authenticity of his work, Ramanujan did his research with the financial support of the head of the Indian Mathematical Society, Ramachandra Rao, who also helped him publish his work in the *Journal of the Indian Mathematical Society.*

As a simple example to demonstrate the genius that Ramanujan possessed, consider the formula he developed to get the value of π:

$$\frac{1}{\pi} = \frac{\sqrt{8}}{9801} \sum_{n=0}^{\infty} \frac{(4n)!}{(n!)^4} \times \frac{26390n + 1103}{396^{4n}}.$$

Another such effort resulted in the following relationship:

$$\frac{3\pi}{4} = \sum_{k=1}^{\infty} \arctan\left(\frac{2}{k^2}\right).$$

He also did some playful things, such as creating this very unusual magic square:

22	12	18	87
88	17	9	25
10	24	89	16
19	86	23	11

First of all, as with all magic squares, all the rows, columns, and diagonals have the same sum. In this case the sum is 139. However, with this unusual magic square there are also additional sums that total to 139, such as the following:

- the sum of the four center squares,
- the sum of the two center squares in the top row and the two center squares in the bottom row,
- the sum of the two center squares in the left-hand column and the two center squares in the right-hand column.
- the sum of the four corner 2 by 2 squares, and
- the sum of the four corner cells.

By now you must realize the genius of this man to create this fabulous magic square. It should also be noted that Ramanujan developed a series of other such magic squares; the topic surely fascinated him.

We might also look at some of his other elementary discoveries, such as his nest of radicals:

$$3 = \sqrt{9}$$

$$= \sqrt{1+8}$$

$$= \sqrt{1+2\cdot 4}$$

$$= \sqrt{1+2\sqrt{16}}$$

$$= \sqrt{1+2\sqrt{1+15}}$$

$$= \sqrt{1+2\sqrt{1+3\cdot 5}}$$

$$= \sqrt{1+2\sqrt{1+3\sqrt{25}}}$$

$$= \sqrt{1+2\sqrt{1+3\sqrt{1+4\cdot 6}}}$$

$$= \sqrt{1+2\sqrt{1+3\sqrt{1+4\cdot 6}}}$$

$$= \sqrt{1+2\sqrt{1+3\sqrt{1+4\sqrt{36}}}}$$

$$= \sqrt{1+2\sqrt{1+3\sqrt{1+\ldots}}}$$

which eventually will look like

$$3 = \sqrt{1+2\sqrt{1+3\sqrt{1+4\sqrt{1+5\sqrt{1+6\sqrt{1+7\sqrt{1+8\sqrt{1+9\sqrt{1+\cdots}}}}}}}}}$$

Most of Ramanujan's discoveries are clearly far beyond the scope of this book, so we present merely a few to demonstrate the brilliance of the man.

Mathematicians in India became fascinated with Ramanujan's talent and began to connect him with mathematicians in England. Some of the English mathematicians did not even reply to his letters because he claimed to have no formal education. However, in 1913, enthralled by the book *Orders of Infinity*, and looking to expand his horizons, Ramanujan wrote a letter to the book's author, the famous English mathematician G. H. Hardy (1877–1947), who was a professor at the University of Cambridge. Once again, he indicated his lack of formal education, but included a collection of some of his findings to see what Hardy would think about them. Hardy was amazed at the ingenuity of what he found in this letter. Relationships were proposed, which he had never seen before, such as

$$\cfrac{e^{\frac{2\pi}{5}}}{1+\cfrac{e^{-2\pi}}{1+\cfrac{e^{-4\pi}}{1+\cdots}}} = \sqrt{\frac{5+\sqrt{5}}{2}} - \phi, \text{ where } \phi \text{ represents the golden radio.}$$

This ultimately prompted Hardy to facilitate Ramanujan's first trip to London in 1914. As the collaboration with Hardy and colleagues began—and they were truly fascinated by his lack of formal education—they were genuinely amazed at his discoveries.

This was about the time when World War I emerged, and food rationing was a problem for Ramanujan, who then began having further health problems. Despite his physical frailty, Ramanujan was invited to enroll as a student at Cambridge University, and by 1916 was awarded a degree, which was then called a bachelor's degree, but today is valued as a PhD. Sadly, his illness became worse in 1917, and through most of the year he spent his time in nursing homes. The following year, he was elected as a fellow of the Cambridge Philosophical Society, and shortly thereafter he received the greatest honor of his life, being elected a fellow of the Royal Society of London, which was formally confirmed in May 1918. This was followed by further acceptance of his brilliance by being elected a fellow of Trinity College of Cambridge University.

The following year Ramanujan returned to India with a reputation that, as it was stated at the time, was greater than any Indian had previously enjoyed. However, his health did not improve much and eventually began to decline even further. He died in India on April 26, 1920, at the age of 32.

Fortunately, much of his writings have been saved and ultimately published. This began shortly after his death when his brother Tirunsrayanan began to collect his handwritten notes for further publication. It should be noted that much of Ramanujan's work was not accompanied by proof, yet it was always shown to be correct. It is speculated that since paper was expensive, he did the proof on a slate and then copied the result on paper. It was not uncommon that slates were used in India at that time.

Posthumously, Ramanujan has received countless accolades. In India, December 22, his birthday, is often celebrated as "National Mathematics Day."

In 1962 the government of India issued a postage stamp with Ramanujan's picture (fig. 43.2). A second stamp (fig. 43.3) was issued in 2011 with a different design, once again commemorating Ramanujan's genius. To further honor his remarkable achievements, the Indian government declared that Srinivasa Ramanujan's birthday, December 22, would be considered National Mathematics Day in perpetuity.

Figure 43.2.

Figure 43.3.

CHAPTER 44

John von Neumann:
Hungarian-American (1903–1957)

Before the advent of modern electronic calculators and computers, all mathematical calculations had to be carried out by humans using paper and pencil. Slide rules or some more sophisticated mechanical devices represented the only technological assistance. It's quite hard to imagine what the world would look like today were this still the case. Did you know that the word "computer" originally referred to a person who carried out calculations or computations? In fact, the word continued with the same meaning until the middle of the twentieth century. Its meaning gradually changed as technological advances led to more and more efficient machines, eventually causing the extinction of "human computers." Times have changed, and nowadays, prodigies in mental calculation are more likely to be seen on a television show than in a research center. However, in the nineteenth and early twentieth century, research agencies such as the National Advisory Committee for Aeronautics (NACA, founded in 1915 and then in 1958 becoming the newly founded National Aeronautics and Space Administration—NASA) still relied on human computers and, naturally, they scouted the highest-performance individuals in this profession. In particular, mental calculators were in great demand. Although there is no comparable job profile today, there are still people around who practice mental calculation very intensively. Every two years, the world's best mental calculators are invited to compete for the Mental Calculation World Cup, which was first held in 2004 in Germany. It is a common misbelief that mathematicians are particularly good in mental calculation. As a matter of fact, many mathematicians even like to emphasize that they have problems with mental arithmetic. Although such statements often come with a pinch of coquetry, there is some truth behind it. A mathematician does not have to be good at mental arithmetic. People with limited mathematical education or interest often think that mathematics is "all about computing numbers," which is not at all the case. Being an excellent mathematician but lousy in mental arithmetic is actually no contradiction. Of course, some outstanding mathematicians are, or were, also exceptional mental calculators aside. The American mathematician John von Neumann, one of the greatest mathematicians of the twentieth century, was a child prodigy in language, memorization, and mathematics. At the age of six, he could divide one eight-digit number by another in his head. Mental calculation was such a natural amusement to him that when he caught his mother staring aimlessly, he asked her: "What are you calculating?" He also possessed an amazing memory. A brief glance at a page of the telephone directory sufficed for him to memorize all its names and numbers. He made major contributions to a number of fields and published more

Figure 44.1. John von Neumann.

than 150 papers in his life. During World War II, von Neumann worked on the Manhattan Project, where he developed the mathematical models that were behind the explosive lenses and worked out key steps in the nuclear physics involved in the hydrogen bomb. By the way, the Manhattan Project is also the most famous modern example for the employment of "human computers" on a massive scale, and it should be mentioned that most of them were women.

John von Neumann was born János Neumann on December 28, 1903, in Budapest, which was then part of the Austro-Hungarian Empire. He was the eldest of three brothers. His father, Miksa (Max) Neumann, was a successful banker and held a doctorate in law. John's mother came from a wealthy Jewish family. However, the family did not observe strict religious practices. In 1913, Emperor Franz Joseph elevated Max Neumann to nobility for his contribution to the then successful economy. His son later used the German form von Neumann where the "von" indicated the nobility title. Until the age of ten, John and his brothers were taught by various governesses, since at that time formal education did not begin earlier in Hungary. John was an exceptional case of a child prodigy. At the age of six, he was able to converse with his father in classical Greek and he showed an amazing memory. The Neumann family sometimes entertained guests with demonstrations of John's memory by letting a guest select a random page of the phone book. After reading over it a few times, young John had memorized the names, addresses, and numbers; he could answer any question put to him. Later he was able to recite whole books such as Goethe's *Faust*. By the age of eight, he was familiar with differential and integral calculus. However, he was particularly interested in history, and by reading a large number of books, he acquired an incredible historical knowledge before entering school. In 1911 von Neumann entered the Lutheran Gymnasium, which was one of the best schools in Budapest. At that time, Hungary had an excellent education system, which produced several outstanding mathematicians and physicists. Among the brilliant and creative minds who were educated in Budapest from their childhoods to their teens are Leó Szilárd (1898–1964), Eugene Wigner (1902–1995), Edward Teller (1908–2003), Paul Erdős (1913–1996), and Peter Lax (1926–), to name just a few. The concentration of great mathematicians who were educated in Budapest in the early twentieth century was so strong that Peter Lax once said, "You don't have to be Hungarian to be a mathematician, but it helps." Apart from the excellent school system, several other factors may have contributed to the emergence of this phenomenon, most notably the Eötvös Mathematics Competition (now renamed as Kürschák Competition), mathematical contests that, since 1894, have been open to Hungarian high school students in their senior year. This is the oldest modern mathematics competition in the world; the problems to solve focused on creativity and not on memorized knowledge. The ten best scorers were exempt from the very competitive entrance examinations at the university. Von Neumann's exceptional giftedness was immediately recognized by his mathematics teacher, who organized private tutoring for him to promote his mathematical talent. At the age of 15, he was sent to the renowned mathematician Gábor Szegő (1895–1985) to study advanced calculus. Szegő was deeply impressed by von Neumann's genius, and subsequently visited the von Neumann house twice a week to tutor the child prodigy. When von Neumann completed his school education, he was already collaborating with professional mathematicians. In spite of his very promising perspectives for a career as a mathematician in academia, his father did not want him to study mathematics. Academic positions for mathematicians in Hungary were not well paid and Max Neumann wanted his son to follow a more financially rewarding path in business or industry. The compromise on which they agreed was that he would study chemistry to become a chemical engineer in industry. Since von Neumann didn't know very much about chemistry, he first took two-year non-degree courses in chemistry at the University of Berlin, after which he sat for and passed the entrance exam to the prestigious ETH Zurich. However, at the same time, von Neumann also entered Pázmány Péter University in Budapest as a PhD candidate in mathematics. Although he couldn't attend any courses in Budapest, he achieved outstanding results in the examinations. He graduated as a chemical engineer from ETH Zurich in 1926 and simultaneously finished his PhD thesis in mathematics, seemingly without appreciable effort. He was then granted a fellowship from the Rockefeller Foundation to study mathematics under David Hilbert in Göttingen. He completed his habilitation in 1927 and became the youngest privat-docent in the history of the University of Berlin in 1928. Von Neumann published highly original mathematical papers at an incredible rate, quickly making him famous in the mathematical community and turning him into a star at academic conferences. By the end of 1929 he had already published thirty-two major papers. After a brief stay in Hamburg in the same year, he accepted an offer from Princeton University in New Jersey. Before moving to the United States, he married Marietta Kövesi in Budapest. After a few years of teaching at Princeton University, he became a mathematics professor at the newly founded Institute for Advanced Study at Princeton. Von Neumann anglicized his first name to John, keeping the German-aristocratic surname of von Neumann. During his first years in the United States, von Neumann continued to return to Europe during the summers and even kept academic posts in Germany, which he resigned when the Nazis came to power. In 1935, Marietta gave birth to a daughter, Marina. Two years later the couple divorced and in 1938, von Neumann married Klara Dan, whom he had met in Budapest during one of his European visits. At Princeton, von Neumann's lifestyle was all but typical for a top mathematician. He enjoyed eating and drinking in company, and almost once a week, John and Klara gave a party at their home, thereby creating a kind of salon. Von Neumann loved jokes, especially Yiddish and "off-color" humor. His memory helped him to always have a joke ready, if a conversation got stuck.

While most mathematicians need a quiet environment for working and studying, von Neumann preferred noisy and chaotic environments. At Princeton he received complaints for regularly playing extremely loud German march music on his gramophone, disturbing colleagues in neighboring offices, including Albert Einstein. He often worked in the couple's living room with the television playing loudly and even read books while driving his car. In combination with the fact that he was a rather reckless driver anyway, this led to frequent traffic tickets and car accidents. Reportedly, an intersection in Princeton was nicknamed "Von Neumann Corner" because of the many car accidents he had there.[1] Despite his unconventional personality, he was generally regarded as the leading mathematician of his time. His genius and brilliant intuition for new mathematical theories enabled him to contribute seminal and path-breaking work in completely different branches of mathematics, as well as in theoretical physics and computer science. Von Neumann applied new mathematical methods to quantum theory and was the first to establish a rigorous mathematical framework for quantum mechanics, known as the Dirac–von Neumann axioms. He founded the field of game theory as a mathematical discipline. Game theory is the study of mathematical models of strategic interaction between rational decision makers, and is now used in economics, political science, philosophy, and computer science. Von Neumann's inspiration for game theory was poker, a game he played occasionally. He realized that a mathematical model for poker that only uses probability theory is insufficient for a thorough analysis of the game, since players' strategies are completely ignored. In particular, he wanted to formalize the idea of "bluffing," a strategy that is meant to mislead other players and hide information from them. The beginning of his mathematical investigation of games can be marked with his 1928 article, "Theory of Parlor Games," which can also be regarded as the initiation of modern game theory. In this article, he proved the so-called minimax theorem. The minimax theorem states that in zero-sum games in which players know at each time all moves that have taken place so far, there exists a pair of strategies for both players that allows each to minimize his maximum losses, hence the name minimax. (In a zero-sum game, each participant's gain or loss of utility is exactly balanced by the losses or gains of the utility of the other participants.) Von Neumann continued his work in game theory, improving and extending his results to include more general games with more than two players. He soon noticed that the mathematical framework he was developing could become an important tool in economics. Consequently, he collaborated with the Austrian economist Oskar Morgenstern (1902–1977), a professor at Princeton University, with whom he published the paper *Theory of Games and Economic Behavior* in 1944. This groundbreaking text is one hundred pages long and established the interdisciplinary research field of game theory. When it became a book, it attracted also public interest. It is a classic foundational work that still belongs to the standard literature in mathematical economics. Moreover, it may also serve as valuable reading for anyone planning a career as a professional poker player.[2] In the late 1930s, von Neumann began to study the mathematical modeling of explosions and soon became the leading authority in this field. This led to frequent military consultancies, and in 1943, von Neumann was invited to work on the Manhattan Project. He made principal contributions to the implosion design of the atomic bomb, enabling for a more efficient weapon.

Von Neumann was also a pioneer in the development of modern computing. After examining the army's ENIAC (Electronic Numerical Integrator and Computer) during the war, von Neumann used his mathematical abilities to improve the computer's logic design. He proposed a new design that embodied the "stored-program" concept that is now called the Von Neumann architecture. He was a consultant in the construction of the army's EDVAC (Electronic Discrete Variable Automatic Computer), one of the earliest binary stored-program computers. In fact, he wrote in ink one of the first programs for EDVAC, a sorting algorithm, covering twenty-three pages. His wife, Klara, became one of the first computer programmers. Von Neumann was one the most influential mathematicians who ever lived. Edward Teller wrote that "Nobody knows all science, not even von Neumann did. But as for mathematics, he contributed to every part of it except number theory and topology. That is, I think, something unique." Other mathematicians were stunned by von Neumann's mental calculation abilities and his incredible speed. The Hungarian-American mathematician Paul Halmos (1916–2006) recounts a story told by physicist Nicholas Metropolis (1915–1999), concerning the speed of von Neumann's calculations, when somebody asked von Neumann to solve the famous fly puzzle:

> Two bicyclists start 20 miles apart and head toward each other, each going at a steady rate of 10 mph. At the same time a fly that travels at a steady 15 mph starts from the front wheel of the southbound bicycle and flies to the front wheel of the northbound one, then turns around and flies to the front wheel of the southbound one again, and continues in this manner till he is crushed between the two front wheels. Question: what total distance did the fly cover? The slow way to find the answer is to calculate what distance the fly covers on the first, southbound, leg of the trip, then on the second, northbound, leg, then on the third, etc., etc., and, finally, to sum the infinite series so obtained.

The quick way is to observe that the bicycles meet exactly one hour after their start, so that the fly had just an hour for his travels; the answer must therefore be 15 miles.

When the question was put to von Neumann, he solved it in an instant, and thereby disappointed the questioner: "Oh, you must have heard the trick before!" "What trick?" asked von Neumann; "All I did was sum the geometric series." That meant doing it the long way—instantly!

After the war, von Neumann served on the General Advisory Committee of the US Atomic Energy Commission, and later as one of its commissioners. He was a consultant to a number of organizations, including the US Air Force, the US Army's Ballistic Research Laboratory, the Armed Forces Special Weapons Project, and the Lawrence Livermore National Laboratory. In 1955, von Neumann was diagnosed with cancer. He died at age 53 on February 8, 1957, at the Walter Reed Army Medical Center in Washington,

DC, under military security lest he reveal military secrets while heavily medicated. His mathematical legacy is perhaps best described in Peter Lax's foreword to von Neumann's *Selected Letters*, edited by Miklós Rédei: "To gain a measure of von Neumann's achievements, consider that had he lived a normal span of years, he would certainly have been a recipient of a Nobel Prize in economics. And if there were Nobel Prizes in computer science and mathematics, he would have been honored by these, too. So the writer of these letters should be thought of as a triple Nobel laureate or, possibly, a 3½-fold winner, for his work in physics, in particular, quantum mechanics."

CHAPTER 45

Kurt Gödel:
Austrian-American (1906–1978)

In the late nineteenth century, Georg Cantor developed set theory, which became a fundamental theory in mathematics. It offered a common foundation to all fields of mathematics, and mathematicians attempted to formalize Cantor's set theory by finding a minimal of axioms, from which all further mathematical statements within the theory can be derived. Although this endeavor seemed very promising at the beginning, the axiomatization of set theory ran into serious problems when it was discovered that it suffered from logical paradoxes and inconsistencies. This led to a severe foundational crisis of mathematics. In response to this crisis, mathematician David Hilbert initiated a program to find a complete and finite set of axioms that would provide a stable basis for all existing mathematical systems, from arithmetic and geometry to advanced calculus and all other fields as well. More complicated systems would be proved in terms of simpler systems and these by even simpler systems, and ultimately the consistency of all mathematics would be reduced to basic arithmetic. More precisely, Hilbert's program to establish secure foundations for all mathematics comprised the following goals:[1]

- A formulation of all mathematics: all mathematical statements should be written in a precise formal language and manipulated according to well-defined rules.

- Completeness: a proof that all true mathematical statements can be proved in the formalism.
- Consistency: a proof that no contradiction can be obtained in the formalism of mathematics.
- Conservation: a proof that any result about "real objects" obtained using reasoning about "ideal objects" (such as uncountable sets, for instance the set of real numbers) can be proved without using ideal objects.
- Decidability: there should be an algorithm for deciding the truth or falsity of any mathematical statement.

Many well-known logicians and mathematicians spent years on such a program, including Alfred North Whitehead (1861–1947) and Bertrand Russell (1872–1970), who published their monumental work as *Principia Mathematica* in three volumes in 1910, 1912, and 1913. However, in 1931, young Austrian mathematician Kurt Gödel proved that the goals that Hilbert had formulated were impossible to achieve. He published two famous theorems, known as Gödel's incompleteness theorems, which ended all attempts to find an all-encompassing set of axioms for mathematics or other formal systems. Roughly speaking, he showed that it is impossible to find a set of axioms sufficient for all mathematics, since for any set of axioms proposed to encapsulate mathematics, either the system must be inconsistent, or there must be some truths of mathematics, which could not be deduced from them. In fact, it is impossible to come up with an axiomatic mathematical theory that captures even all of the truths about the natural numbers. Gödel's results had a deep impact on the foundations of mathematics, as well as on logic and philosophy. Their significance is perhaps best described by John von Neumann, who said that "Kurt Gödel's achievement in modern logic is singular and monumental—indeed it is more than a monument, it is a landmark which will remain visible far in space and time. . . . The subject of logic has certainly completely changed its nature and possibilities with Gödel's achievement."

Kurt Gödel was born on April 28, 1906, in Brno (German: Brünn), which is now the second largest city in the Czech Republic, but then was part of the Austro-Hungarian Empire. Before World War I, the majority of the population of Brno was German speaking. Gödel's family was rather wealthy; his father, Rudolf Gödel, was the managing director of a textile factory in Brno. He was Catholic and his wife, Marianne, was Protestant.

Figure 45.1. Kurt Gödel in 1925.

Kurt and his elder brother, Rudolf, were raised Protestant in a country with a Catholic majority. Kurt went through several episodes of poor health as a child. When he was six years old, he suffered from rheumatic fever. Although he recovered well, he became convinced that his heart was permanently damaged as a result of the illness. He arrived at this conclusion when he began to read medical books at the age of 8, initiating a lifelong hypochondria. When the Austro-Hungarian Empire broke up at the end of World War I, Czechoslovakia declared its independence and Gödel's family automatically became Czechoslovak citizens, suddenly belonging to a German-speaking minority in the Republic of Czechoslovakia. However, Gödel could barely speak Czech and felt alien in this newly founded state. It was common that many of the German-speaking residents still considered themselves Austrian. By the time Gödel completed his school education in Brno, he had mastered university mathematics. Besides mathematics, languages were his favorite subjects. His brother later recalled that during his whole high school career, Kurt had made not a single grammatical error in Latin; needless to say, his schoolwork had always received the top marks. In 1923, Gödel took Austrian citizenship and moved to Vienna. He entered the University of Vienna as a student of theoretical physics. Among his teachers were physicist and philosopher Moritz Schlick (1882–1936) and mathematicians Hans Hahn (1879–1934), Karl Menger (1902–1985), and Philipp Furtwängler (1869–1940). Furtwängler was paralyzed from the neck down and lectured from a wheelchair with an assistant who wrote on the board. He was a brilliant mathematician and his lectures had a big influence on Gödel, making him change his major subject to mathematics. A seminar on Bertrand Russell's book *Introduction to Mathematical Philosophy*, conducted by Moritz Schlick, awakened his interest in mathematical logic. Already as a student, Gödel participated in the Vienna Circle (German: Wiener Kreis), which was a group of philosophers and scientists drawn from the natural and social sciences, logic and mathematics, who met regularly from 1924 to 1936 at the University of Vienna, chaired by Schlick. Through the Vienna Circle Gödel learned about David Hilbert's program and the foundational crisis of mathematics. He attended a lecture by Hilbert in Bologna on completeness and consistency of mathematical systems and chose this topic for his doctoral work. In 1929, at the age of twenty-three, he completed his doctoral dissertation under the supervision of Hans Hahn. He was awarded his doctorate in 1930 and the Vienna Academy of Science published his thesis. In 1931 Gödel published his famous article *Über formal unentscheidbare Sätze der "Principia Mathematica" und verwandter Systeme* ("On Formally Undecidable Propositions of 'Principia Mathematica' and Related Systems"), containing his incompleteness theorems. Today, almost all relevant mathematics journals accept only articles written in English. However, before World War II there also existed quite a few top journals in other languages, notably in German. Gödel's article originally appeared in German in the journal *Monatshefte für Mathematik*, which was founded in 1890 in Austria. The journal still exists and is published by Springer in cooperation with the Austrian Mathematical Society; although it has kept its German name, all articles are now in English. In his 1931 article, Gödel proved that for any computable axiomatic system that is powerful enough to describe the arithmetic of the natural numbers, the following is true:

1. If a (logical or axiomatic formal) system is consistent, it cannot be complete.
2. The consistency of axioms cannot be proved within their own system.

Gödel proved that it is impossible to use the axiomatic method to construct a mathematical theory that entails all of the truths of mathematics. The incompleteness theorem was an extremely important negative result and had a huge impact in the field of mathematical logic. To prove this theorem, Gödel invented a new method, now known as Gödel numbering. He assigned to each symbol and well-formed formula of the theory a unique natural number (now called Gödel number). Moreover, he showed that classical paradoxes of self-reference, such as "This statement is false," can be recast as self-referential formal sentences of arithmetic. Gödel's incompleteness theorem ended a half-century of attempts to find a set of axioms sufficient for all mathematics. Showing that it is impossible to construct axiomatic systems that could be used to prove all mathematical truths, he destroyed a whole branch of research. Gödel's result quickly made him famous and invitations to International Mathematical Congresses followed. In 1933, he made his first trip to the United States, where he met Albert Einstein and delivered an address to the annual meeting of the American Mathematical Society. In the same year, Hitler came to power in Germany and over the following years, the Nazis' influence grew in Austria as well. With the rise of the Nazis in Austria, many of the Vienna Circle's members left for the United States and the United Kingdom. During these years, Gödel traveled a lot and gave lectures at the newly founded Institute for Advanced Study in Princeton. However, he kept his base in Austria. When Moritz Schlick was murdered by one of his former students in 1933, Gödel suffered a severe nervous breakdown and spent several months in a sanatorium. In addition to his hypochondria, he developed a fear of being poisoned, and showed other symptoms of paranoia. In 1938, Austria became a part of Nazi Germany. Gödel, who had been privat-docent at the University of Vienna, had to renew his application under the new order, but the University of Vienna turned it down. His former association with Jewish members of the Vienna Circle might have played a role in the decision. In autumn of that year, Gödel married Adele Porkert. They had been in a relationship for several years, but Gödel's parents had objected to a marriage. She was six years older than Gödel and had been married before. Moreover, she was not highly educated and was Catholic, while Gödel was Protestant. When World War II started in September 1939, Gödel feared that he might be conscripted into the German army. The couple left Vienna for Princeton, where Gödel accepted a position at the Institute for Advanced Study. They had to take the Trans-Siberian Railway to the Pacific and sail from

Japan to San Francisco. At Princeton, Gödel developed a close friendship with Albert Einstein, and they often took long walks around the Institute together (see fig. 45.2). In 1947, Einstein and the Austrian economist Oskar Morgenstern, also at Princeton, accompanied Gödel to his US citizenship exam. In his preparation for the exam, Gödel believed he had found an inconsistency in the US Constitution and Einstein was worried that his friend would explain his discovery to the judge and thereby spoil his application. Fortunately, the judge knew Einstein, and everything went well. In 1949, Gödel discovered an exact solution of the Einstein field equations, which is also known as the Gödel universe. It is a solution describing a rotating universe and among other unusual properties it would allow time travel into the past. However, the solution is somewhat artificial in that the so-called cosmological constant, a parameter of the theory, which is now associated with dark energy, has to be fine-tuned to a very specific value. Astronomical observations at that time could neither exclude nor confirm that we live in a rotating universe. As observational data continually improved, Gödel would also continue to ask astronomers, until his death, "Is the universe

Figure 45.2. Kurt Gödel and Albert Einstein in Princeton.

rotating yet?" and be told "No, it isn't." For his work in relativity, Gödel was awarded (with physicist Julian Schwinger) the first Albert Einstein Award in 1951. Gödel remained in Princeton for the rest of his life. A permanent member of the Institute for Advanced Study at Princeton since 1946, he became a full professor in 1953. During his first years in the United States, he continued to publish fundamental mathematical papers. However, in his later years, he devoted more and more time to studying philosophy. He admired the works of Leibniz and began writing about philosophical issues. As Gödel aged, his paranoia got worse. His fear of being poisoned became obsessive, and he would eat only food that his wife, Adele, prepared for him. But in 1977 she suffered a stroke and was hospitalized for several months. During this time, she had to watch her husband continuously losing weight because he refused to eat. When she left the hospital, Gödel weighed only 66 pounds, whereupon she immediately brought him to the hospital. Further treatment was too late, and he died a few weeks later, on January 14, 1978. He essentially starved to death. His death certificate reported that he died of "malnutrition and inanition caused by personality disturbance." His wife, Adele, died in 1981. Gödel's incompleteness theorems and some of his other mathematical works are ranked among the greatest mathematical achievements of the twentieth century. He was one of the most significant logicians in history. His name is also known from a popular 1979 book, *Gödel, Escher, Bach* by Douglas Hofstadter. The book won the Pulitzer Prize for general nonfiction and the National Book Award for Science. It explores relationships between the works of Gödel, along with those of artist M. C. Escher and composer Johann Sebastian Bach.

~

CHAPTER 46

Alan Turing:
English (1912–1954)

It was widely believed that the work of a brilliant mathematician shortened World War II by two years and perhaps saved as many as 14 million lives. The immediate question would be how can a mathematician do something that is typically left for soldiers to accomplish? The answer is that through his unique genius, the English mathematician Alan Turing developed a machine that was able to break the Nazi enigma code during World War II, and thereby discover their strategic plans in advance.

In September 1938, Turing began working at the Government Code and Cypher School (GC&CS), which was an organization that specialized in breaking war codes. During World War II the GC&CS was located at Bletchley Park, Milton Keynes, England—which today is a museum largely celebrating Turing's work. Turing's main responsibility at the time was cryptanalysis of the Enigma, which was a ciphering system developed by the Western Allies in World War II to decipher Morse-coded radio communications of the Axis powers that had been enciphered through Enigma machines (see fig. 46.1). Essentially, through his innovative work, we say that Alan Turing was the founder of computer science as we know it today.

Figure 46.1. Enigma Machine.

In figure 46.2 we show instructions as to how the machine functioned.

Initially, the Allies used a Polish system to decipher the Axis war codes, but in time it proved to be ineffectual; the Germans were able to supersede its findings. Turing came up with an improvement that made the enigma

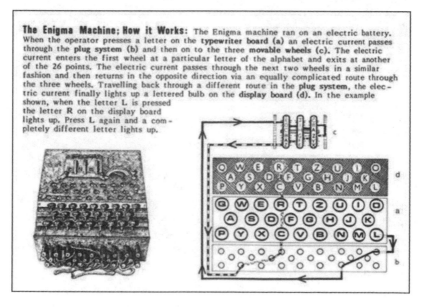

The Enigma Machine: How it Works: The Enigma machine ran on an electric battery. When the operator presses a letter on the **typewriter board (a)** an electric current passes through the **plug system (b)** and then on to the three **movable wheels (c)**. The electric current enters the first wheel at a particular letter of the alphabet and exits at another of the 26 points. The electric current passes through the next two wheels in a similar fashion and then returns in the opposite direction via an equally complicated route through the three wheels. Travelling back through a different route in the **plug system**, the electric current finally lights up a lettered bulb on the **display board (d)**. In the example shown, when the letter L is pressed the letter R on the display board lights up. Press L again and a completely different letter lights up.

Figure 46.2.

machine effective. Essentially, he was able to break intercepted codes that enabled the Allies to defeat the Germans in key battles such as the Battle of the Atlantic. His discoveries for deciphering codes were considered an act of sheer brilliance.

Alan Mathison Turing was born in Paddington, London, England, on June 23, 1912, to a well-placed British family. Since his parents were civil servants working in India, Turing spent many of his early years in British foster homes, where there was little encouragement academically. At age 6 he was enrolled in St. Michael's day school, then at age 10 the Hazel Hurst Preparatory School. His mother was very concerned about his education and wanted him to enter a private school, which he did successfully by entering the prestigious all-boys Sherborne School—which was founded in 1550. There he showed a greater interest in mathematics and science than in the classics, which caused the headmaster there to caution his parents about his potentially failing this primary subject of the school. He managed to grapple with all the subjects, but particularly showed an incredible talent for solving mathematics problems before age 16, and not yet having been exposed to elementary calculus.

Figure 46.3. Alan Mathison Turing.

Turing began his undergraduate studies in 1931 at King's College of Cambridge University, where for the first time he felt he had a real home. It is also the time where his homosexuality became an integral part of his existence. Rather than spend much of his time in literary circles, he spent a great deal of time in physical activities such as rowing and running. Later as a working mathematician, he would often run to work as many as 10 miles each way and often beating his colleagues who would have taken public transportation at the same time. Further to his running skill it should be noted that in 1948 he ran a marathon in 2 hours 46 minutes 3 seconds, a time that was only 11 minutes slower than the Olympic champion that year!

In 1934, he graduated from King's College with honors, whereupon he was elected a fellow of King's College. He began publishing rather impressive mathematical papers, one of which reworked the Austrian mathematician Kurt Gödel's universal arithmetic-based language with simple hypothetical devices that became known as the Turing machine. This was an abstract machine that manipulated symbols on a strip of tape according to a table of rules. Given any computer algorithm, a Turing machine could be constructed that would be capable of simulating that algorithm's logic. In his 1948 essay, "Intelligent Machinery," Turing wrote that his machine consisted of

> an unlimited memory capacity obtained in the form of an infinite tape marked out into squares, on each of which a symbol could be printed. At any moment there is one symbol in the machine; it is called the scanned symbol. The machine can alter the scanned symbol, and its behaviour is in part determined by that symbol, but the symbols on the tape elsewhere do not affect the behaviour of the machine. However, the tape can be moved back and forth through the machine, this being one of the elementary operations of the machine. Any symbol on the tape may therefore eventually have an innings (or lifespan).[1]

As mentioned earlier, this was the beginning of today's computer. The famous Hungarian-American mathematician John von Neumann (see chap. 44) further supported that idea by saying that the central concept of the modern computer was due to Turing's work.[2]

Turing then began his doctoral studies in 1936 at Cambridge University and eventually obtained his PhD degree in 1938 from Princeton University, whereupon John von Neumann encouraged him to stay on as a postdoctoral assistant. However, Turing decided to return to England instead. Upon his return in September 1938, and after attending lectures at Cambridge University, he began working part time at the GC&CS. However, the day after Britain declared war on Germany on September 3, 1949, Turing began to work full time at the GC&CS, where his work eventually brought him lasting fame. There, Turing helped develop an electromechanical machine that could break the enigma code more effectively than the previously used Polish machine, *Bomba Kryptologiczna*, which in its new form was referred to as *Bombe* (see fig. 46.4). Yet the most important aspect was Turing's ingenious mechanization of subtle logical deductions. They were able to routinely read Luftwaffe signals, but not the German naval communications. Once again it was Turing who was eventually able to break these previously unbreakable codes before the end of 1941.

By 1942, Turing was considered the genius of Bletchley Park, and also had a rather wanting physical appearance, coupled by halting speech and

Figure 46.4. A complete and working replica of
a *Bombe* at the National Codes Centre at Bletchley Park.

somewhat strange behavior. One of his colleagues was Joan Clarke, who seemed to have caught his eye and to whom he proposed marriage. This was enthusiastically accepted; however, shortly thereafter Turing withdrew his offer and exposed to her his homosexuality. This would not have stopped her from marrying him, but he could no longer go forward in that regard.

In November 1942 he came to the United States to further work on the U-boat Enigma crisis, which had still perplexed the Allies; they couldn't decipher the signals. By March 1943 the U-boat enigma decryption was effectively resolved and remained so for the rest of the war. Once again, his brilliance became a key factor supporting the Allied troops in the war.

In the later years of the war, Turing had worked with electronic enciphering speech in the telephone system. Although this eventually produced successful results, they were mostly too late to be useful during the war. After the war, Turing lived in London, where he worked on the automatic computing engine, which was a significant forerunner to today's computers. Secrecy still permeated the field and a lot of his contributions to the development of computers was not publicized until after his death. Beginning in 1948, Turing held the position of reader in the mathematics department at Victoria University in Manchester, England, where he also worked at the computing machine laboratory and helped develop software for the earliest stored-program computer called the *Manchester Mark 1*. There, he also dabbled with the notion of artificial intelligence, which was one of the earliest attempts to see how a computer could correspond with a human being. In a rather indirect way, these early results by Turing are used today on the internet when we wish to find out if the user is a human or a computer. This test is called CAPTCHA.

In 1948 Turing turned his attention to work with colleagues to develop a program for a computer that would allow it to play chess against a human being. Eventually this became successful, but the moves by the computer took as much as a half-hour; it did in fact beat some competitors, but not all.

By 1951 Turing took up an interest in biology, albeit from a mathematical point of view. He was fascinated by how biological organisms develop their shape. For example, he wanted to understand how phyllotaxis seemed to be dominated by the Fibonacci numbers.[3] In more general terms he studied morphogenesis. His work in this field is still relevant today as a defining portion of mathematical biology. Turing's work in biology has helped understand the growth of organisms that determine placement of feathers and hair follicles, as well as location of various parts of the human body.

In January 1952, Turing developed an intimate relationship with a British man, who was 20 years his junior. Soon thereafter, it was discovered that Turing had a sexual relationship with this man, which at the time in the United Kingdom was a criminal offense. At the trial, Turing's attorney did not dispute the offense, and entered a guilty plea. To avoid imprisonment, he accepted an alternative, which was to undergo hormonal treatment to reduce his libido. The treatment caused him to be impotent and produced unpleasant bodily changes. His conviction also removed his security clearance for his cryptographic work.

On June 8, 1954, he was found dead in his home, which later was determined to be a result of cyanide poisoning and deemed a suicide. It was 55 years later, in 2009, as a result of extensive petitions, that Britain's Prime Minister Gordon Brown acknowledged the inappropriateness of condemning Turing's homosexuality as criminal and offered an apology. This, however, did not satisfy many of those who felt that Turing's treatment was unjustified and counterproductive to our scientific advancement. As a result of many years of petitions and attempts in Parliament, on December 24, 2013, Queen Elizabeth II signed a pardon for Turing's conviction for gross indecency, as it was then called.

Figure 46.5.

Today Alan Turing is heralded as the father of computers and the initiator of investigations in a number of areas of science and mathematics. Although he was appointed to the Order of the British Empire in 1946, and then elected as a Fellow of the Royal Society in 1951, his name appears in countless mathematical and scientific concepts and university buildings and halls throughout the world. This is further evidenced by the fact that he was chosen from a list of 227,299 nominees—including Charles Babbage and Ada Lovelace—to be pictured on the highest-denomination banknote in England, the fifty-pound banknote, as shown in figure 46.5. For those interested in this unusually brilliant person, Alan Turing's life is chronicled in the 2014 movie *The Imitation Game*.

~

Chapter 47

Paul Erdős:
Hungarian (1913–1996)

It is not uncommon that people who have a true genius mentality are often socially unusual. There is probably no better example of this than the Hungarian mathematician Paul Erdős, who had no stable residence, traveled endlessly visiting one mathematician after another while carrying all his belongings in one suitcase. It was clear that all he cared about was mathematics—making conjectures and proving them as theorems. He connected with most of the world-famous mathematicians of his lifetime. However, curiously enough, he published profusely and very often with other mathematicians as coauthors. One of the many legacies that he left behind is what is known today as the *Erdős number*, which is assigned to mathematicians as follows: If a mathematician coauthored an article with him, he or she had an Erdős number 1. If a mathematician coauthored an article with another mathematician who already had an Erdős number of 1, then he or she was assigned an Erdős number 2. If a mathematician coauthored an article with a mathematician who already had an Erdős number 2, then he or she would be assigned an Erdős number 3, and so it would continue. Of course, Paul Erdős himself had Erdős number 0. In other words, there is great prestige in mathematical circles of having any Erdős number at all. As a matter of fact, Albert Einstein had an Erdős number 2. Incidentally, the American Mathematical Society provides a free online tool to compute the Erdős number of an author.[1] Nowhere else in the mathematics world does this kind of jubilation take place.

Figure 47.1. Paul Erdős.

Paul Erdős spent almost all of his waking hours doing mathematics either by himself or in conversation with others. Oftentimes, he spent eighteen hours a day engrossed with mathematics. It is believed that he collaborated with more than 500 mathematicians and wrote more than 1,500 mathematical papers during his lifetime. This is likely one of the largest productions of mathematics contributions in history.[2]

Paul Erdős was born in Budapest, Hungary on March 26, 1913, to parents who both were high school mathematics teachers. He was particularly treasured by his parents, since his two sisters died of scarlet fever in their youth on the day of his birth. His childhood began in a rather strange fashion, since his father was a prisoner of war in Siberia till 1920, and so to support the family his mother needed to leave him alone at home, where he entertained himself by looking at mathematics books that were lying around the house. And at a very early age he showed an incredible facility for doing mathematical calculations in his head, such as multiplying two three-digit numbers in his head at age three. When his father returned from Siberia, he recognized his son's talents and began to move them along so that he started to embrace such topics as number theory, infinite series, combinatorics, and set theory.

By the age of seventeen, Erdős entered the Pázmány Péter Catholic University of Budapest, where he had already began publishing articles, such as a proof of Chebyshev's theorem, which states that for any integer $n > 3$, there always exists at least one prime number p, where $n < p < 2n - 2$, were stated another way, if $n > 1$, then there is always at least one prime number p that is between n and $2n$.

By age twenty-one, he completed his undergraduate work and received his doctorate in mathematics. With the rise of anti-Semitism in Hungary in 1934, Erdős decided to leave the country and begin a four-year postdoctoral fellowship at the University of Manchester in England. In 1938, he accepted a one-year appointment at the Institute for Advanced Study at Princeton. One of his great achievements there was to participate in the development of the field of probabilistic number theory. After his stay at Princeton he began to travel around the United States, visiting Purdue University, Stanford University, Notre Dame University, and Johns Hopkins University, while turning down full-time positions at any one of these, so that he could work with a multitude of mathematicians of his choice and when he wanted to. This was the beginning of his nomadic travels that took him to more than 25 countries throughout the world. He had no family to concern himself with, and stayed with mathematicians as long as the challenges remained of interest to him. Often time, he would show up unannounced with suitcase in hand, prepared to stay as long as he or the host wished. He often worked more than eighteen hours a day and it is believed he was oftentimes stimulated by various medications to keep him alert.

In 1949, along with the Norwegian-American mathematician Atle Selberg (1917–2007), who had just begun his career as a professor of mathematics at the Institute for Advanced Study, Princeton, Erdős reached a point of jubilation when they proved the prime number theorem, which is a formula that gives the approximate value of the number of prime numbers less than or equal to any given positive real number x, which is usually notated as $\pi(x)$.[3] In other words, $\pi(2) = 1$—that is, there is only one prime less than or equal to 2—namely, the number 2 itself. As another example, $\pi(10) = 4$, which represents the following prime numbers less than or equal to 10, namely, 2, 3, 5, and 7. In general terms, the prime number theorem states that for larger values of x,

$$\pi(x) \approx \frac{x}{\ln x}.\text{[4]}$$

In figure 47.2, we show the number of primes less than or equal to n, for selected values.

For this work and other discoveries in prime number theory, Erdős was awarded the Cole Prize, which was presented to him by the Hungarian-American mathematician John von Neumann (see chap. 44) in 1951. Through the next several decades, Erdős seemed to concentrate his work on combinatorics, number theory, set theory, and geometry, just to name a few to demonstrate the broad spectrum of interest that dominated his life. When it came to graph theory, he helped organize the first international conference in 1959. As a further example of his unusual value system, living on very little money and using his relationship with the multitude of mathematicians that he visited to subsist, in 1984 he was awarded the prestigious (and lucrative) Wolf prize worth $50,000, of which he kept only $720 and gave the remainder to establish a scholarship program in Israel in memory of his parents.

Despite his odd lifestyle, Erdős was revered around the world. He received more than fifteen honorary degrees and has been elected as a member of the scientific academies of eight countries, including the US National Academy of Sciences and the British Royal Society. In his own way, he also gave back to society by offering payments to mathematicians who solved previously unsolved problems with an amount depending upon the difficulty of the problem considered—ranging from $25 to several thousand dollars for very difficult problems. One problem that had truly challenged Erdős and still awaits a solution is referred to often as the Collatz conjecture, which was first discovered in 1932 by the German mathematician Lothar Collatz (1910–1990), who then published it in 1937. This is one problem for which Erdős offered $500 for a proof.

To appreciate this conjecture, we begin by following two rules as using an *arbitrarily* selected number.

n	$\pi(n)$ **The number of primes less than or equal to n**	$\dfrac{\pi(n)}{n}$ **The proportion of primes among the first n numbers**
10^2	25	0.2500
10^4	1,229	0.1229
10^6	78,498	0.0785
10^8	5,761,455	0.0570
10^{10}	455,052,511	0.0455
10^{12}	37,607,912,018	0.0377

Figure 47.2.

If the number is *odd,* then multiply by 3 and add 1.
If the number is *even,* then divide by 2.

Regardless of the number selected, after continued repetition of the process, it is conjectured that we will always end up with the number 1.

Let's try it for the *arbitrarily selected* number 7:

7 is odd, therefore, multiply by 3 and add 1 to get: $7 \cdot 3 + 1 = \mathbf{22}$

22 is even, so we simply divide by 2 to get **11**

11 is odd, so we multiply by 3 and add 1 to get **34**.

34 is even, so we divide by 2 to get **17**.

17 is odd, so we multiply by 3 and add 1 to get **52**.

52 is even, so we divide by 2 to get **26**.

26 is even, so we divide by 2 to get **13**.

13 is odd, so we multiply by 3 and add 1 to get **40**.

40 is even, so we divide by 2 to get **20**.

20 is even, therefore, divide by 2 to get **10**.

10 is even, therefore, divide by 2 to get **5**.

5 is odd, so we multiply by 3 and add 1 to get **16**.

16 is even, so we divide by 2 to get **8**.

8 is even, so we divide by 2 to get **4**.

4 is even, so we divide by 2 to get **2**.

2 is also even, so we again divide by 2 to get **1**.

If we were to continue, we would find ourselves in a loop (that is, 1 is odd, so we multiply by 3 and add 1 to get **4**, . . .). After 16 steps, we end up with a 1, that, if we continue the process, would lead us back to 4, and then on to the 1 again. We end up in a loop! Therefore, we get the sequence:

7, 22, 11, 34, 17, 52, 26, 13, 40, 20, 10, 5, 16, 8, **4, 2, 1** , **4, 2, 1** , . . .

The following schematic (fig. 47.3) will show the path we have just taken:

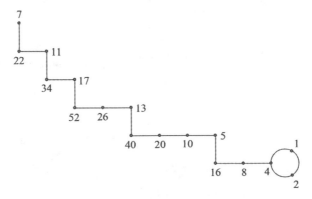

Figure 47.3. For arbitrary number *n* = 7.

It is also interesting to see a graph (fig. 47.4) of the steps of this process.

Regardless of which number we begin with (here we started with **7**), we will eventually get to 1.

This is truly remarkable! Try it for some other numbers to convince yourself that it really does work. Had we started with **9** as our arbitrarily selected number, it would have required **19** steps to reach 1. Starting with **41** will require 109 steps to reach 1.

Paul Erdős lived a full and apparently satisfied life completely engrossed with mathematics. He died at the age of eighty-three on September 20, 1996, of a heart attack while attending a mathematics conference in Warsaw, Poland. He was buried in a grave next to that of his parents in Budapest, and for his epitaph he offered, "I've finally stopped getting dumber."

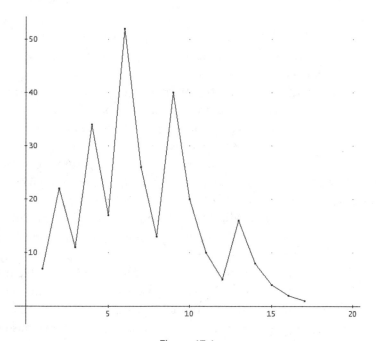

Figure 47.4.

CHAPTER 48

∼

Herbert A. Hauptman:
American (1917–2011)

The mathematician Herbert Aaron Hauptman enjoys the distinction of being the first mathematician to win the Nobel Prize—albeit the prize was for chemistry, since there is no Nobel Prize category set aside for mathematics. There are many stories as to why Alfred Nobel chose not to include mathematics among the categories of his prize. Yet the most common one is that there was a jealousy involving a woman and a competing mathematician. There are many variations and other rumors regarding reasons that there was no such prize in mathematics. However, after Hauptman had won the prize in 1985, other mathematicians have since been awarded a Nobel Prize, such as John Forbes Nash Jr. (1994) and Robert J. Aumann (2005).

Dr. Hauptman won the Nobel Prize for his work thirty years earlier where he solved the phase problem of X-ray crystallography. Actually, he had used mathematical methods to solve a forty-year-old problem that chemists were unable to solve. When he produced his results in 1955, he was severely criticized because it was felt that this problem was not solvable. The technique that evolved from his research has had an enormous effect on our pharmaceutical industry. The process that results from his work, which he called "shake and bake," allows the pharmaceutical researchers to determine the crystal structure of the bad germs so that they can then construct the appropriate medicine to combat them. As Dr. Hauptman has often said, if one were to use the fly and mosquito killer of the 1950s, known as Flit, the insects today would simply laugh and crawl away, since, over

Figure 48.1. Herbert Aaron Hauptman.

time, they have developed an immunity against this spray. The same, he said, occurs in the bacteria or virus world, where they build an immunity against combative pharmaceuticals. The system he helped develop allows the pharmaceutical industry to continue to develop new and effective drugs through an analysis of the crystal structure of the bacteria or other harmful cells, and therefore, develop appropriate combatants.

It is interesting to view the story as to how Dr. Hauptman reached this exclusive position in his career. He was born in the Bronx, New York, on February 14, 1917. He attended the local public schools, where he excelled in mathematics and won entry into the most prestigious high school in the country at the time, Townsend Harris High School, which admitted boys through a very challenging entrance examination. This was a three-year high school with an automatic admission to the City College of New York, at the time a highly sought-after tuition-free college, which to date has had ten of its former students winning the Nobel Prize—more than any other public institution in the United States.

Prior to his graduation from City College, he was awarded the very prestigious Belden Prize in Mathematics in 1936 and graduated with a BS in mathematics in 1937. Because the United States was then in a severe depression, jobs were very difficult to come by. However, as a mathematics major, there were always jobs available in New York City as a teacher of mathematics. At the time, there were several exams one had to pass to qualify for the position of high school teacher. One of these exams was a speech test. Fortunately, or unfortunately, Hauptman failed the speech test, since he was told that he had a Bronx dialect, which at that time was unacceptable. Thereupon, he entered Columbia University, earning a master of arts degree in mathematics in 1939. With the war now in full swing, Hauptman enlisted in the navy, where he served as a weather forecaster in the South Pacific.

After the war, he decided to seek an advanced degree and pursue a career in basic scientific research, whereas teaching was no longer an option. There he entered into a partnership with Jerome Karle, a chemist who graduated from City College the same year as Hauptman, although, interestingly enough, they did not know each other during their student years. While he worked at the Naval Research Laboratory in Washington, DC, he simultaneously enrolled in the PhD program at the University of Maryland. And so began a multiyear collaboration of the mathematician Dr. Hauptman and a physical chemist, Dr. Karle. Their 1953 monograph, "Solution of the Phase Problem I—The Centrosymmetric Crystal," which relied heavily on Dr. Hauptman's mathematical talents, contains the main ideas of his research, the most important of which was the introduction of the joint probability distributions of several structure factors as the essential tool for phase determination. In this monograph, they also introduced the concepts of the structure invariants and semi-invariants, special linear combinations of the phases, and used them to devise recipes for origin specification in all the centrosymmetric space groups. The notion of the structure invariants and semi-invariants proved to be of particular importance because they also served to link the observed diffraction intensities with the needed phases of the structure factors.

With a clear picture of the structure of hormones and other biological molecules, researchers better understood the chemistry of the body and of drugs used to treat various illnesses. For example, once they understood the structure of enkephalins, pain-control substances found naturally in the body, they were able to make progress in developing new pain-killing drugs.

It must be said that the mathematical talent that Dr. Hauptman provided to the chemistry field enabled him to produce results that had the additional benefit of greatly speeding up the analysis of molecular structures. In the 1960s, it could take two years to work out the structure of a simple antibiotic molecule that had only fifteen atoms. Through his discoveries, it became possible to determine the structure of a fifty-atom molecule in two days.

Dr. Hauptman, a mathematician, was not inducted into the National Academy of Sciences–chemistry section, at the time when Dr. Karle was inducted, largely because he was not a chemist. Since there had never been an American selected for the Nobel Prize in a science who was not previously a member of the National Academy of Sciences, there was little chance that he would ever get the Nobel Prize. They were wrong! As soon as he was announced as a Nobel laureate (for chemistry), he was quickly invited and encouraged to join the National Academy of Sciences–chemistry section! Thus, Dr. Hauptman was the first Nobel Prize winner in a science who was not previously a member of the National Academy of Sciences.

One of Dr. Hauptman's hobbies was to determine how to optimally pack with various-sized marbles the contents of polyhedra. He even published his findings; this became a studied branch of geometry (see fig. 48.2).

Figure 48.2.

Figure 48.3.

By 1970, he joined the crystallography group of the Medical Foundation of Buffalo, which is today known as the Hauptman-Woodward Medical Research Institute, where he served as president until he died in Buffalo, New York, on October 23, 2011. There he continued his innovative research to further improve the field of crystallography and enhance the pharmaceutical industry's ability to create new and effective medications in a very efficient fashion. So here we have another example where brilliance in mathematics allows the sciences to flourish.

It must be said that Hauptman was a truly wonderful person, who had very few idiosyncrasies that one typically finds among genius personalities. It was well known that he never liked to comb his hair at a time when that was still fashionable, so his wife would do that every day for him as you can see in a photograph (fig. 48.3) with one of the authors (Posamentier) taken in 2008. It didn't disturb him that his watch was always twelve minutes too slow; he could easily calculate the correct time. In a book that he coauthored with Posamentier, he was so proud to have developed a factorial function $r!$ for rational, nonintegral values of r.[1] Although research showed that this might have been anticipated by Gauss, he was delighted that he was in such good company. Suffice it to say, he was a truly wonderful person— universally loved by all!

CHAPTER 49

~

Benoit Mandelbrot:
Polish-American (1924–2010)

There are times when a mathematician is largely known for one mathematical discovery. This is the case with the mathematician Benoit Mandelbrot, who was born in Warsaw, Poland, on November 20, 1924, although through his peripatetic life he has held both French and American citizenships. He was always fascinated with geometry. Even as a boy it is said that he saw chess games rather geometrical than logical. Later in life through his innovative publication *The Fractal Geometry of Nature*,[1] he asks, "Why is geometry often described as cold and dry? One reason lies in its inability to describe the shape of a cloud, a mountain, a coastline or a tree." His primary claim to fame within the realm of mathematics is his development of *fractals*, a field in geometry comprised of objects in similar patterns with increasingly smaller scales. We will inspect fractals in greater detail after we consider the lifestyle of Mandelbrot, which brought him to these curious discoveries.

Benoit Mandelbrot spent the first eleven years of his life in Poland in a family that was rather academic, his mother being a dentist. However, it was two of his uncles who introduced and motivated him toward mathematics. In 1936, with the rise of Nazism, his family emigrated to France where his uncle, who was a professor of mathematics, took responsibility for Mandelbrot's education. Studying in Paris at the start of World War II was rather difficult and gave him an opportunity to think about mathematics independently, which allowed him to gravitate further toward geometry. In an

Figure 49.1. Benoit Mandelbrot.

attempt to avoid the Nazis, who occupied much of France at the time, he left Paris with his family and continued his studies in Tulle, France. In 1944 he returned to Paris to continue his studies at the Lycée du Parc on Lyon, and from 1945 to 1947 he attended the École Polytechnique. From there he went on to the California Institute of Technology, where, in 1949, he received a master's degree in aeronautics. He then went back to France to the University of Paris, where he earned a doctorate in mathematics in 1952. Soon thereafter, he left Paris to return once again to the United States, this time to the Institute of Advanced Study at Princeton, where he was mentored by John von Neumann. Once again on the move, in 1955 he went to France to work at the Centre National de la Recherche Scientific, where he met and married Aliette Kagan, and soon the couple moved to Switzerland and then again back to France. Finally, the couple, still mobile, moved back to the United States where he took the position of a research fellow at the IBM Thomas J. Watson Research Center in Yorktown Heights, New York, because Mandelbrot felt uncomfortable with the French style of mathematics study, whereas the IBM environment allowed him greater freedom in exploring mathematics from his geometrical viewpoint. He remained at IBM for the next thirty-five years.

In 1980, with the aid of the computer, Mandelbrot showed that pictures of mathematical objects created in 1918 by the French mathematician Gaston Julia (1893–1978) were quite beautiful, not monstrous as some may have felt. And more importantly, he showed that, rather than pathological, the ragged outlines and the repeating patterns of those figures were often found in nature. (See fig. 49.2 for some examples.) Mandelbrot used the Latin word *fractus*, meaning broken or fractured, to coin a word to denote the new mathematical objects: *fractals*.

In figure 49.2 the images to the left are pictures of real-life scenes. Those pictures to the right are related fractal models. The characteristic feature of fractals is self-similarity: Geometric patterns seen in the big picture

Figure 49.2.

of a fractal are repeated in their parts in smaller and smaller scales. Making fractals involves the repeated application of a geometric rule, or transformation, of an original figure or set of points, which we will refer to as the *seed* of the fractal.

Once we determine what the generative procedure and the seed of a fractal will consist of, we can begin constructing the fractal by repeatedly applying the generative procedure—first to the seed, then once again to the resulting output, and so on. There lies another definitive aspect of fractal construction: It is made up of consecutive phases called *iterations*. An iteration is the act of applying one algorithm or procedure one time through in a repetitive process.

When constructing a fractal, the iterations of the generative procedure are done recursively—that is, the input of each iteration is the output of the previous one—with the exception of the first iteration, which is applied to a seed. In some cases, this means that each subsequent iteration will be more cumbersome than the previous one. In those cases, programmable technology is definitely immensely helpful.

A fractal ideally entails the iteration of a procedure an infinite number of times, although in practice we can iterate a procedure only a finite number of times. We can use computers to help us perform as many iterations as we want, which would give us different stages in the construction of a fractal. Or we can use mathematics to deduce what would be the result of performing that infinite process. Let us consider the generative process described above, and the appropriate terminology through a classical example, the Koch Snowflake (fig. 49.3).[2]

For the construction of this fractal, the *seed* will be an equilateral triangle. Because the generation of the fractal happens by successive iterations, we will call the result of each iteration a *stage* in the fractal construction. The *generative procedure* will consist of erasing the middle third of every line segment (the initiator) in a stage and replacing it with two line segments of the same length (one-third of the length of the original segment) at an angle of 60°; this will form cusps (which look like partial equilateral triangles—the generator), where before there were segments. We can see this procedure illustrated in figure 49.3.

Each *iteration* will consist of applying the fractal construction procedure to each line segment in a stage of the fractal, which will create the next stage. Figure 49.4 shows the first two iterations in the construction of the Koch Snowflake.

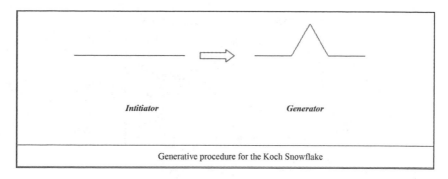

Figure 49.3. Generative procedure for the Koch Snowflake.

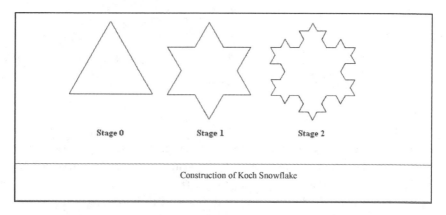

Figure 49.4. Construction of Koch Snowflake.

We can sketch stage 3 of the Koch Snowflake on a separate piece of paper, or use a computer drawing program. Remember this iteration will require applying the generative procedure to every segment in stage 2. It involves quite a bit more work than the previous iteration. While the first iteration consisted of applying the generative procedure to three line segments, in the second iteration that number increased to twelve. In the third iteration we will need to work on forty-eight line segments. That increase in complexity is displayed in figure 49.5. With each iteration, each line segment at a stage will be replaced by four line segments, forming a cusp, in the next stage. So, if we know how many segments there are at a stage, we can find out the number of segments at the next stage by multiplying that number by four. This relation can be written algebraically and recursively as: S_n

Stage	Number of Line Segments
n	(S_n)
0	3
1	12
2	48
3	192
n	$4 \times S_{n-1}$

Figure 49.5.

$= 4 \times S_{n-1}$ (which just happens to equal 3×4^n), where S_n is the number of segments at stage n, and S_{n-1} is the number of segments at the previous stage.

The replacement of each segment by a visual spike, and the accelerating increase in the number of segments in this fractal, is what gives it the main features of a fractal: the jagged appearance and the property of self-similarity. If we zoom in on any spike, we will find smaller and smaller copies of it. Magnifying fractals reveals in them small-scale details similar to the large-scale characteristics.

Another popular fractal is the Sierpiński Gasket (fig. 49.6).[3] Its seed is also an equilateral triangle. Each iteration consists of splitting a triangle into four smaller equilateral triangles by using the midpoints of the three sides of the original triangle as the new vertices, then deleting the middle triangle

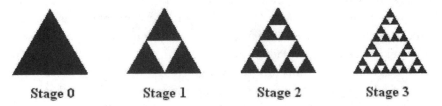

Figure 49.6. Construction of the Sierpiński Gasket.

from further action (a quarter of the area is deleted). The construction of the fractal continues by iterating this procedure over and over: From every new triangle formed, a triangle is removed from its interior. This results in not only a rough, fragmented surface, but also in self-similarity—the two main features of fractals.

The Fibonacci sequence has made its way into one of the most well-known fractals: the Mandelbrot set (fig. 49.7). But first, let's see what the Mandelbrot set is. Its image is so popular it could earn the title of the "emblem of fractal geometry." Its strange beauty mesmerizes laypeople and experts alike. But what does that image represent? As with the other fractals we have examined, some elements are involved in its construction: a seed, a rule or transformation, and an infinite number of iterations. But unlike our previous examples—which were mainly geometrical—the Mandelbrot set is a set of numbers. The image we see in figure 49.7 is just a plot, in the complex plane,[4] of the numbers that belong to the set.

How do we know whether a number is or is not in the Mandelbrot set? We must test each number to find out. This infinitely large task can only be

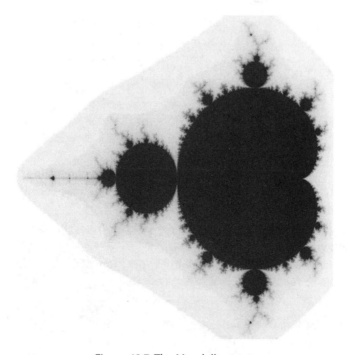

Figure 49.7. The Mandelbrot set.

done with the aid of a computer, and only a finite number of times, although a very large number of times. In fact, it was only under the right conditions, in which Benoit Mandelbrot's vision and intellect was combined with the environment of IBM's Watson Research Center, that a revival of work on this set that had been initiated by Julia in the 1920s was made possible.

So the construction of the image of the Mandelbrot set requires one more element besides the seed, the rule, and the iterations that the fractals, previously discussed, also had: It involves a test of numbers. Let us say the number we are testing is c.

The seed for this fractal is the number zero; not a triangle or a segment, but a number, because this fractal is numerical in nature. The rule or transformation is: "square the input and add c," which can be expressed algebraically as $x^2 + c$.

Suppose we want to test the number $c = 1$. Our transformation becomes: $x^2 + 1$.

Let us see the result of a few iterations, starting with the seed 0 as the input, and then using the output of each iteration as the input for the next:

$$0^2 + 1 = 1$$
$$1^2 + 1 = 2$$
$$2^2 + 1 = 5$$
$$\vdots$$
$$5^2 + 1 = 26$$
$$\vdots$$
$$26^2 + 1 = 677$$
$$\vdots$$
$$677^2 + 1 = 458{,}330$$

We can see that with more iterations, the greater the result will be. The terms of the sequence of numbers will increase without bound. We say that "it goes to infinity."

Let us test for another number, $c = 0$. With this value for c, our rule becomes: $x^2 + 0$.

Starting with the same seed 0, a few iterations will show that the sequence will be fixed at zero:

First iteration: $0^2 + 0 = 0$
Second iteration: $0^2 + 0 = 0$.

As a third example, let us take $c = -2$, for which the rule becomes: $x^2 - 2$. We start with the same seed -2 and again obtain a fixed sequence after the first iteration:

First iteration: $(-2)^2 + (-2) = 4 - 2 = 2$
Second iteration: $2^2   + (-2) = 4 - 2 = 2$

For each value of c, the "test" (repeatedly iterating the rule) will tell us whether the result will go to infinity, or if it will not. Values of c that will result in an escape to infinity are *not* in the set; all the others *are* in the set. The image of the Mandelbrot set is actually a record of the fate of each number, c, under this test.[5] The key to understanding the image is to unveil the code used. The most frequently used code for plotting the results of these tests is to use the color black to represent those points in the plane that *are* in the Mandelbrot set, and to color the others according to their "escape speed"—that is, using different colors to represent the number of iterations that value takes to reach a certain distance from the origin. Another traditional way of plotting the Mandelbrot set is just to use black for points that are in the set and white for those that are not.

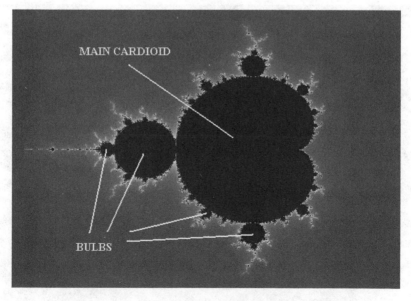

Figure 49.8. The main cardioid and bulbs in the Mandelbrot set.

We will now look at the image of the Mandelbrot set with a categorical eye. At the core of the image, we can see a heart-shaped figure, the *main cardioid.*[6] We can also note many round decorations, or *bulbs* (fig. 49.8). We call any bulb that is directly attached to the main cardioid a *primary bulb.* The primary bulbs have in turn many smaller decorations attached to them. Among them, we can identify what appear to be *antennas* (fig. 49.9).

We will call the longest of these antennas the *main antenna.* Finally, the main antennas show several "spokes" (fig. 49.10). Note that the number of spokes in a main antenna varies from decoration to decoration. We will call this number the *period* of that bulb or decoration. To determine the period of that bulb, just count the number of spokes on an antenna. We must remember to count the spoke emanating from the primary decoration to the main junction point. Figure 49.11 displays various primary bulbs and their periods.

How can the Fibonacci sequence be seen in the Mandelbrot set? We will consider the period of the main cardioid to be 1. Then, by counting the spokes in the main antenna of the largest primary bulbs, we will determine their period. The result of this counting—that is, the period of the main

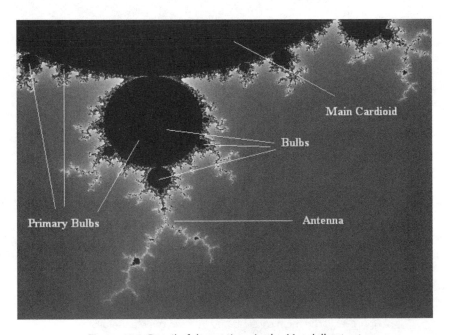

Figure 49.9. Detail of decorations in the Mandelbrot set.

Figure 49.10. Main antennas and their "spokes."

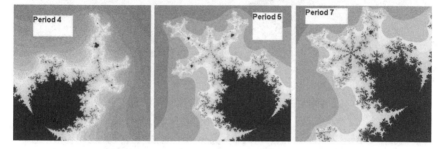

Figure 49.11. Determining the period of bulbs by counting
the "spokes" in their main antenna.

cardioid and of some of the largest primary bulbs, are registered in figure 49.12.

It is surprising to see from an inspection of figure 49.12 that the largest bulb between the bulb of period 1 and the bulb of period 2 is a bulb with period 3. The largest bulb between the period-2 bulb and a period-3 bulb is a period-5 bulb. And the largest bulb between a period-5 bulb and a period-3 bulb is a period-8 bulb. Interestingly, the Fibonacci numbers seem to appear. There are no obvious explanations as to why they appear. The Fibonacci numbers are not related directly to the way in which the periods of primary bulbs are calculated. The Fibonacci sequence inexplicably makes a mysterious and remarkable appearance—just another striking feature of fractals.

As well as being an IBM Fellow at the Watson Research Center, Mandelbrot held numerous other academic positions such as Professor of the Practice of Mathematics at Harvard University, Professor of Engineering

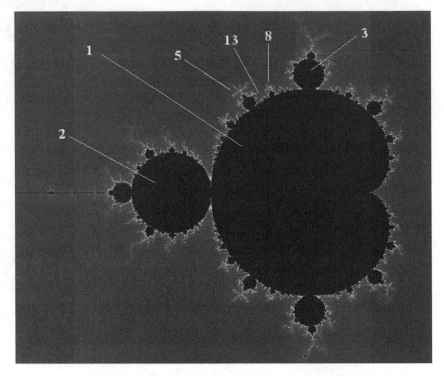

Figure 49.12. The Fibonacci numbers in the Mandelbrot set.

at Yale, Professor of Mathematics at the École Polytechnique, Professor of Economics at Harvard, and Professor of Physiology at the Einstein College of Medicine in New York. Mandelbrot's excursions into so many different branches of science were intentional. It was, however, the fact that fractals were so widely found that in many cases, they provided the route into other areas. Mandelbrot also received a multitude of academic honors and prizes, most of which he received in between the years of 1985 and 2003, such as in 1985 he received the Barnard Medal for Meritorious Service to Science, in 1986 the Franklin Medal, in 1987 the Alexander von Humboldt Prize, in 1988 the Steinmetz Medal, in 1989 the Légion d'Honneur and the Nevada Medal, in 1993, the Wolf Prize for Physics, and in 2003 the Japan Prize for Science and Technology, as well as others.

On October 14, 2010, Mandelbrot died of pancreatic cancer in Cambridge, Massachusetts. He was lauded universally not only for his development of fractals, but for his universal intelligence. His obituary in *The Economist* highlights his fame as "celebrity beyond the academy."[7]

CHAPTER 50

Maryam Mirzakhani: Iranian (1977–2017)

It has always been a puzzle to determine exactly why there is no Nobel Prize for mathematics. Speculations abound as to why Alfred Nobel chose not to designate mathematics as one of his prize categories, but nothing conclusive has emerged. This is not to say that mathematicians cannot win the Nobel Prize. A case in point is Herbert A. Hauptman. He was the first mathematician to win the Nobel Prize, albeit for chemistry, as he solved a 40-year-old chemistry problem using his mathematical talent (see chap. 48). There is, however, a prize solely for mathematicians under age 40. That is the Fields Medal, which is issued every four years to either two, three, or four mathematicians who have exhibited extraordinary genius. It is awarded by the International Mathematics Union. The first woman to win this prestigious award was the Iranian mathematician Maryam Mirzakhani, in 2014. Officially she won the award for her outstanding contributions to the dynamics and geometry of Riemann surfaces and their moduli spaces.

Sadly, life was cut short at age forty due to a severe case of cancer, while she was a professor of mathematics at Stanford University in California. Dr. Mirzakhani's life began on May 12, 1977, in the Iranian city of Tehran. Perhaps her initial motivation for the subject came from her father, who was an electrical engineer. Her talents were recognized early as she attended a school for gifted youngsters, the Tehran Farzanegan School. In 1994 at age seventeen she won a gold medal at the International Mathematical Olympiad, becoming the first Iranian female student to win such an honor.

Figure 50.1. Maryam Mirzakhani.

This was followed by the next year's International Mathematical Olympiad, where she achieved a perfect score, winning two medals and once again carrying the distinction of being the first Iranian student holding this honor.

From there she went on to Sharif University of Technology to earn her baccalaureate degree in 1999. She then attended Harvard University, earning her PhD in 2004. Once again, her superb talents enabled her to become a research fellow at the Clay Mathematics Institute, as well as holding a professorship at Princeton University. Her family and colleagues often mused about her style of problem-solving mathematics, which involved little diagrams or portals surrounded by mathematical formulas. In short, she had a very peculiar style of gently approaching problems in great depth.

In 2008, Mirzakhani married a Czech mathematician, Jan Vondrak, who was a professor at Stanford University, and then joined him on the faculty as a professor in 2009. As indicated earlier, she died prematurely in California on July 14, 2017, at the age of forty. However, she received a multitude of accolades from her home country, Iran, led by then-president Hassan Rouhani. In 2017, she was elected posthumously to the American Academy of Arts and Sciences.

What she left behind that will solidify her fame in the future are various contributions to the theory of moduli spaces of Riemannian surfaces.

Even a very basic introduction to this branch of mathematics would go far beyond the scope of this book. That is why we will try, instead, to convey at least a very rough idea of the mathematical questions Mirzakhani studied and the methods she used by illustrating the most relevant notions with a few simple examples, without giving precise definitions and omitting technical details. A Riemannian surface can be imagined as a (two-dimensional) surface in space as those shown in figure 50.2: On the left we show a plane—that is, a "flat" surface with zero curvature; then we have a curved surface, where the curvature may vary across the surface; on the bottom we have a sphere, which is a closed surface with constant curvature (the curvature is the same at each point on the surface).

However, not every surface is a Riemannian surface. A surface is called Riemannian if it possesses some additional properties, one of which is called orientability: A surface is called orientable if a two-dimensional figure that is drawn on the surface cannot be moved around the surface and back to where it started so that it looks like its own mirror image. Two examples of non-orientable surfaces are shown in figure 50.3: the Moebius strip (on the

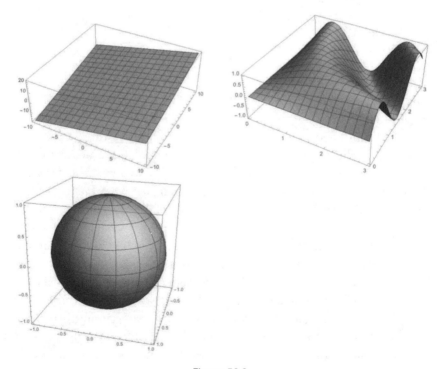

Figure 50.2.

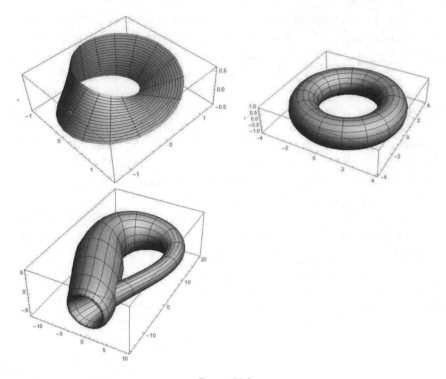

Figure 50.3.

left) and the Klein bottle (below it). On the other hand, a torus (which is the mathematical notion for the shape of a donut) is orientable.

The realm of Riemann surfaces can be divided into three classes: hyperbolic, parabolic, and elliptic Riemann surfaces. These notions correspond to negative curvature, zero curvature (flat), and positive curvature. A sphere is an example of a surface with constant positive curvature, a plane has constant curvature zero, and an example of a hyperbolic surface is a "saddle," shown if figure 50.4.

Hyperbolic surfaces represent the biggest and most diverse group among Riemannian surfaces. Moreover, while elliptic and parabolic surfaces can be further divided into subcategories, no such classification is possible for hyperbolic surfaces. Maryam Mirzakhani's early work was concerned with hyperbolic surfaces, more precisely with closed geodesics on hyperbolic surfaces. A geodesic is a generalization of the notion of a "straight line" to curved surfaces. The term "geodesic" stems from geodesy, the science of

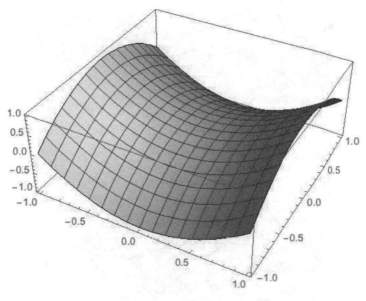

Figure 50.4.

measuring the size and shape of the Earth. Originally, a geodesic meant the shortest route between two points on the Earth's surface, but the abstract mathematical definition of a geodesic as a curve of shortest length also applies to any Riemannian surface, and even to higher-dimensional "surfaces" (also called hyper-surfaces). If we assume the surface of the Earth to be a perfect sphere, then the geodesics are exactly the great circles—that is, circles whose center is at the center of the sphere. Obviously, a sphere has infinitely many closed geodesics (a curve is closed if it has no endpoints). The shape of the Earth is indeed pretty close to that of a sphere, as you can verify by looking at the image shown in figure 50.5, taken by a NASA camera onboard the Deep Space Climate Observatory satellite, one million miles away from the Earth.

Isaac Newton had already discovered that the effect of the rotation of the Earth results in a slight deviation from a spherical shape. The Earth is flattened at the poles and bulges at the equator, resembling a slightly ob-late spheroid (an ellipsoid of revolution). An oblate (flattened) spheroid is obtained if an ellipse is rotated about its minor axis. However, the Earth's deviation from a spherical shape is only about one-third of a percent; dis-tances from points on the surface of the Earth to its center range from 6,353

Figure 50.5.

km (3,948 miles) to 6,384 km (3,965 miles), with a mean radius of 6,371 km (3959 miles). For many practical purposes, such as navigation on the sea, the deviation from a perfect spherical shape can be safely ignored, at least in most circumstances. However, from a mathematical point of view, this "symmetry-breaking" changes the picture completely: for any given point on a sphere, we can find infinitely many closed geodesics running through this point—namely, all the great circles that can be drawn through this point. Yet, the only simple closed geodesics on an oblate spheroid are the meridians (great circles running through the North and South Poles) and the equator (see fig. 50.6a).

In particular, if we take any point on a spheroid that lies not on the equator and is not one of the poles, then there is only one simple closed geodesic running through this point. (*Simple* means that the geodesic closes on itself without an intervening self-intersection.) If we further reduce the symmetry by considering a triaxial ellipsoid (a surface that may be obtained from a spheroid by stretching or compressing it in a direction perpendicular to its axis of rotation; see fig. 50.6b), then we will only find three simple closed geodesic—namely, the equators defined by the three axes of symmetry of the ellipsoid (see fig. 50.7).

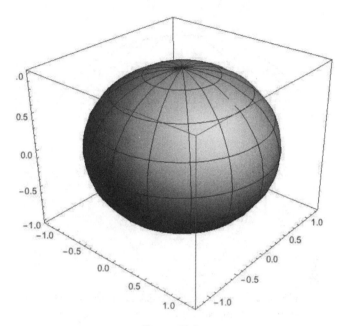

Figure 50.6a.

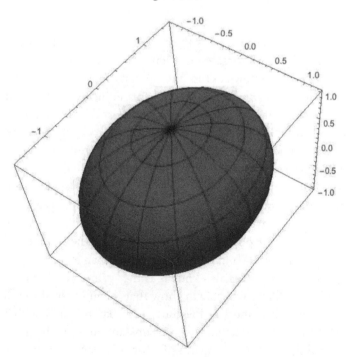

Figure 50.6b.

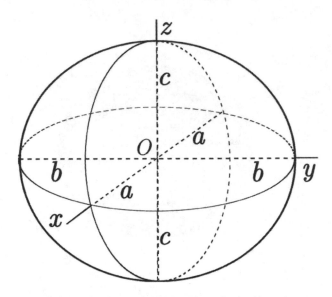

Figure 50.7. Image by Peter Mercator [CC BY-SA 3.0 (https://creativecommons.org/
licenses/by-sa/3.0)]. https://commons.wikimedia.org/wiki/
File:Ellipsoid_tri-axial_abc.svg

As a matter of fact, *three* is the minimum number of simple closed geodesics any Riemann surface must have that is obtained by deforming a sphere. This result is known as the "theorem of three geodesics." It was conjectured by the French mathematician Henri Poincaré in 1905, and finally proved by the German mathematician Hans Werner Ballmann in 1978. We mentioned earlier that Maryam Mirzakhani was especially interested in hyperbolic Riemann surfaces, which constitute the largest and most complex class of Riemann surfaces. It has been known for more than fifty years that on a hyperbolic surface, the number of closed geodesics, whose length is less than some bound L, grows exponentially with L; more precisely, it is asymptotic to $\frac{e^L}{L}$ for large L. This result has a striking similarity to the "prime number theorem" for positive integers, estimating the number of primes less than a given size (the number of primes less than e^L is asymptotic to $\frac{e^L}{L}$ for large L). Therefore, it is also known as the "prime number theorem for geodesics." Mirzakhani was able to show that the number of simple closed geodesics of length at most L does not grow exponentially with L, but is asymptotic to $c \cdot L^{(6g-6)}$, where c is some constant and g is the genus of the surface. Loosely speaking, the genus of a Riemann surface is the number of

Figure 50.8.

holes it has. A sphere has genus 0, a torus or donut has genus 1, and in figure 50.8 we also show a genus-2 surface and a genus-3 surface.

To prove her result on the number of closed and simple geodesics on a hyperbolic surface, Mirzakhani used the concept of the moduli space of all Riemann surfaces with genus *g*. Two Riemann surfaces are said to be topologically equivalent if they can be deformed into each other by continuous deformations. For example, a coffee mug and a torus are both genus-1 surfaces and can be deformed into each other in a continuous fashion (see fig. 50.9)—that is, without any cutting.

This topological equivalence gave rise to a joke among mathematicians, describing a topologist as someone who cannot tell the difference between a coffee mug and a donut. A given topological surface can take on a huge variety of geometric shapes via continuous deformations. For a topological surface of genus *g*, these deformations depend on $(6g) - (6)$ parameters or "moduli." These moduli define by themselves a mathematical space of dimension $(6g) - (6)$ with certain geometric properties, called the moduli space of Riemann surfaces of genus *g*. In her work, Mirzakhani established a link between calculations on abstract moduli space and the counting problem for simple closed geodesics on a single surface, allowing her to translate mathematical results from one world to the other. Not only

Figure 50.9.

could she answer questions regarding geodesics on hyperbolic surfaces by rephrasing them in moduli space, but the connection she had discovered also provided new insights into moduli space. In her highly original proofs, Mirzakhani brought together several mathematical disciplines and built bridges that made powerful mathematical tools of one discipline available to the other. This has led to significant advances in each of them and will also have a considerable influence on their future development.

Epilogue

We hope you have enjoyed our journey through the history of mathematics via the lives of those whom we believe are the most significant mathematicians who have developed the subject to the present day. Selecting fifty outstanding mathematicians from the Western world is a difficult task, and one that is open to alternative assessments. There are clearly many other outstanding mathematicians who could just as easily have been included in our collection; however, we tried to choose those who helped define mathematics as we know it today. On both ends of the spectrum, it is a difficult task to summarize the unusual lives of these mathematicians. For the early days, our resources were very limited. In some cases, there are no written documents available from the highlighted person and so we had to rely on the commentary of other mathematicians who knew of their work. One such example is Thales of Miletus, where most of the information available is a collection of commentaries written by others who flourished during his time and shortly thereafter. On the other end of the spectrum, the difficulty is to describe very advanced mathematics to the general readership, which we have tried to do in the clearest possible way.

It is also noteworthy that those of unusually high intelligence, which we often referred to as genius, have a lifestyle that is not typical of the average citizen. We also notice that these brilliant mathematicians struggled throughout their lives to achieve their groundbreaking ideas and concepts. Oftentimes, they met with resistance and had to grapple with societal issues to publicize their ideas. These included, but were not limited to, poverty, gender, religious beliefs, and other societal peculiarities. Yet these aspects of their lives add further interest as we try to understand how they reached these heights.

We hope that having shed light on these unusually brilliant people will allow the reader an even greater appreciation for mathematics and motivate a desire to pursue further the work of these fifty mathematicians and others that we couldn't fit into this book.

Appendix

Hilbert's Axioms

I. Incidence

1. For every two points A and B there exists a line a, which contains both points. We write $AB = a$, or $BA = a$. Instead of "contains," we may also employ other forms of expression; for example, we may say "A lies upon a," "A is a point of a," "a goes through A and through B," "a joins A to B," etc. If A lies upon a, and at the same time a lies on another line b, we also make use of the expression "The lines a and b have the point A in common."

2. For every two points there exists no more than one line that contains them both; consequently, if $AB = a$, and $AC = a$, where $B \neq C$, then also $BC = a$.

3. There exist at least two points on a line. There exist at least three points that do not lie on the same line.

4. For every three points A, B, C not situated on the same line there exists a plane α that contains all of them. For every plane, there exists a point that lies on it. We write plane $ABC = \alpha$. We may also use the expressions "A, B, C, lie in α," "A, B, C are points of α," etc.

5. For every three points A, B, C that do not lie in the same line, there exists no more than one plane that contains them all.

6. If two points A, B of a line a lie in a plane α, then every point of a lies in α. In this case, we say, "The line a lies in the plane α," etc.

7. If two planes α, β have a point A in common, then they have at least a second point B also in common.

8. There exist at least four points not lying in a plane.

II. Order

1. If a point B lies between points A and C, B is also between C and A, and there exists a line containing the distinct points A, B, C.

2. If A and C are two points, then there exists at least one point, B, on the line AC, such that C lies between A and B.

3. Of any three points situated on a line, there is no more than one that lies between the other two.

4. Pasch's Axiom: Let A, B, C be three points not lying in the same line, and let a be a line lying in the plane ABC and not passing through any of the points A, B, C. Then, if the line a passes through a point of the segment AB, it will also pass through either a point of the segment BC or a point of the segment AC.

III. Congruence

1. If A, B are two points on a line a, and if A' is a point upon the same or another line a', then, upon a given side of A' on the straight line a', we can always find a point B' so that the segment AB is congruent to the segment $A'B'$. Every segment is congruent to itself.

2. If a segment AB is congruent to the segment $A'B'$ and also to the segment $A''B''$, then the segment $A'B'$ is congruent to the segment $A''B''$; that is, if $AB \cong A'B'$ and $AB \cong A''B''$, then $A'B' \cong A''B''$.

3. Let AB and BC be two segments of a line a, which have no points in common, aside from the point B, and, furthermore, let $A'B'$ and $B'C'$ be two segments of the same or of another line a' having, with, likewise, no point in common other than B'. Then, if $AB \cong A'B'$ and $BC \cong B'C'$, we have $AC \cong A'C'$.

4. Let an angle $\angle(h, k)$ be given in the plane α and let a line a' be given in a plane α'. Suppose also that, in the plane α', a definite side of the straight line a' be assigned. Denote by h' a ray of the straight line a' emanating from a point O' of this line. Then in the plane α' there is one, and only one ray k' such that the angle $\angle(h, k)$, or $\angle(k, h)$, is congruent to the angle $\angle(h', k')$, and at the same time all interior points of the angle $\angle(h', k')$ lie upon the given side of a'.

5. If the angle $\angle(h, k)$ is congruent to the angle $\angle(h', k')$ and to the angle $\angle(h'', k'')$, then the angle $\angle(h', k')$ is congruent to the angle $\angle(h'', k'')$.

6. If, in the two triangles ABC and $A'B'C'$, the following congruences are true: $AB \cong A'B'$, $AC \cong A'C'$, $\angle BAC \cong \angle B'A'C'$, then the congruence $\angle ABC \cong \angle A'B'C'$ and $\angle ACB \cong \angle A'C'B'$ also holds true.

IV. Parallels

1. Euclid's Axiom: Let a be any line and a point A not on the line. Then there is at most one line in the plane, determined by a and A, that passes through A and does not intersect a.

V. Continuity

1. Archimedes's Axiom: If AB and CD are any segments then there exists a number n such that n segments CD constructed contiguously from A, along the ray from A through B, will pass beyond the point B.

2. Axiom of line completeness: An extension of a set of points on a line with its order and congruence relations that would preserve the relations existing among the original elements as well as the fundamental properties of line order and congruence that follows from Axioms I–III and from V–1 is impossible.

   Hilbert's 21st Axiom: Any four points A, B, C, D of a line can always be labeled so that B shall lie between A and C and also between A and D, and, furthermore, that C shall lie between A and D and also between B and D. (In 1902 the American mathematician Eliakim Hastings Moore [1862–1932] proved that this 21st axiom was redundant.)

Notes

INTRODUCTION

1. Galileo's book *Il Saggiatore* (1623), see also https://en.wikipedia.org/wiki/The_Assayer.

2. *The Man Who Knew Infinity*. Film written by Matthew Brown and Robert Kanigel (2016).

3. Kanigel, Robert, *The Man Who Knew Infinity: A Life of the Genius Ramanujan,* New York: Washington Square Press (Simon & Schuster), 1991.

CHAPTER 1

1. This is mentioned in the book *The World of Mathematics,* Vol. 1, by James Roy Newman, (New York: Dover Publications, 2000).

CHAPTER 2

1. Alfred S. Posamentier, *The Pythagorean Theorem: The Story of Its Power and Beauty* (Amherst, NY: Prometheus Books, 2010).

2. See, for instance, https://books.google.at/books/about/Pythagoras.html?id=2gLPbFKwY5EC&redir_esc=y.

3. The five Platonic solids are the tetrahedron (a pyramid made of equilateral triangles), the cube (made of six squares), the octahedron (a double pyramid made of eight equilateral triangles), the dodecahedron (made of twelve pentagons), and the icosahedron (made of twenty equilateral triangles).

4. See, for instance, *Stanford Encyclopedia of Philosophy*, https://plato.stanford.edu/entries/pythagoras/.

5. Pythagoras is mentioned in Plato's "The Seventh Letter" according to this book: http://www.sunypress.edu/p-3369-essays-in-ancient-greek-philoso.aspx. The

statement can be found on page 67: Preus, Anthony, ed. *Essays in Ancient Greek Philosophy VI Before Plato*. ISBN13: 978-0-7914-4955-4.

6. Aristotle wrote a monograph titled "On the Pythagoreans," see https://www .jstor.org/stable/283647?seq=1#page_scan_tab_contents.

7. Elisha S. Loomis, *The Pythagorean Proposition*, 2nd ed. (Reston, VA: National Council of Teachers of Mathematics, 1968).

8. James A. Garfield, "Pons Asinorum," *New England Journal of Education* 3 (1876): 116.

CHAPTER 3

1. *Lives of Eminent Philosophers*, edited by Tiziano Dorandi, Cambridge: Cambridge University Press, 2013 (Cambridge Classical Texts and Commentaries, vol. 50, new radically improved critical edition). Translation by R. D. Hicks. (Eudoxus is in Book 8.)

CHAPTER 4

1. *"Classics of Mathematics,"* Ronald Calinger, ed. Oak Park, IL: Moore Publishing, 1982; *Euclid's Elements*, Dana Densmore, ed. Santa Fe, NM: Green Lion Press, 2003; *A History of Mathematics*, V. J. Katz, 3rd ed. New York: Addison-Wesley/ Pearson, 2009.

2. http://www.abrahamlincolnonline.org/lincoln/speeches/autobiog.htm.

3. https://www.nps.gov/liho/learn/historyculture/debate4.htm.

CHAPTER 5

1. No authentic portraits of Archimedes have survived, so the Canadian sculptor R. Tait McKenzie, who designed the medal, had to imagine Archimedes's appearance, inspired by earlier portrayals of Archimedes by Renaissance artists.

2. Chisholm, Hugh, ed. (1911). "Vitruvius," *Encyclopædia Britannica* (11th ed.). Cambridge University Press.

3. His daughter (Marie Louis Sirieix was a man) knew where it was and tried to sell it. Cambridge University only had one single page of palimpsest (the one von Tischendorf had excised), but the whole "book" was in Sirieix's cellar. *The Archimedes Codex: How a Medieval Prayer Book Is Revealing the True Genius of Antiquity's Greatest Scientist*, by Reviel Netz and William Noel, Da Capo Press, 2007.

4. The new owner of the book. According to Simon Finch, who represented the anonymous buyer, stated that the buyer was "a private American" who worked in "the high-tech industry." See https://en.wikipedia.org/wiki/Archimedes_ Palimpsest.

5. This quote is from the chapter "The Life of Marcellus" in the book *The Parallel Lives by Plutarch*. A reproduction of *The Parallel Lives* as published in Vol. V of the Loeb Classical Library edition, 1917, can be found on this webpage: http://penelope.uchicago.edu/Thayer/E/Roman/Texts/Plutarch/Lives/Marcellus*.html. The quote is from page 481.

CHAPTER 6

1. Cyrene was an ancient Greek and later Roman city near present-day Shahhat, Libya. A relatively reliable online source for this is http://www-groups.dcs.st-and.ac.uk/history/Biographies/Eratosthenes.html.

CHAPTER 7

1. The chord of 60 degrees is the length of a line segment whose endpoints are on the unit circle and are separated by 60 degrees.

CHAPTER 8

1. See https://en.wikipedia.org/wiki/Diophantus.

CHAPTER 10

1. Maxey Brooke, "Fibonacci Numbers and Their History through 1900," *Fibonacci Quarterly* 2 (April 1964): 149.

CHAPTER 11

1. Gerolamo Cardano, "A Point of View: Are Tyrants Good for Art?" BBC, August 10, 2012, http://www.bbc.com/news/magazine-19202527.

2. Victor J. Katz, and Karen Hunger Parshall, *Taming the Unknown: A History of Algebra from Antiquity to the Early* (Princeton, NJ: Princeton University Press, 2014); MacTutor History of Mathematics archive, School of Mathematics and Statistics, University of St Andrews, Scotland, link: http://www-groups.dcs.st-and.ac.uk/history/Biographies/Cardan.html. *Encyclopedia Britannica*: https://www.britannica.com/biography/Girolamo-Cardano.

3. "The Story of Mathematics," website by Luke Mastin, link: http://www.storyofmathematics.com/16th_tartaglia.html; MacTutor History of Mathematics archive, School of Mathematics and Statistics, University of St Andrews, Scotland, link: http://www-history.mcs.st-andrews.ac.uk/Biographies/Tartaglia.html.

4. MacTutor History of Mathematics archive, School of Mathematics and Statistics University of St Andrews, Scotland, link: http://www-history.mcs.st-and.ac.uk/HistTopics/Tartaglia_v_Cardan.html; Benjamin Wardhaugh, *How to Read Historical Mathematics* (Princeton, NJ: Princeton University Press, 2010).

5. John Stillwell, *Mathematics and Its History* (Science & Business Media, 2013), Section 5.5, p. 54.

CHAPTER 12

1. The following paragraphs are derived from Alfred S. Posamentier and Bernd Thaller, *Numbers: Their Tales, Types, and Treasures* (Amherst, NY: Prometheus Books, 2015), pp. 212–20.

CHAPTER 13

1. William J. Broad, "After 400 Years, A Challenge to Kepler: He Fabricated His Data, Scholar Says," *New York Times*, Science Section, January 23, 1990, p. 1.

2. The English mathematician Thomas Simpson re-discovered Kepler's rule one hundred years after Kepler. However, he also developed more elaborate approximation formulas, generalizing the formula Kepler had found.

3. A frustum of a cone is the remaining part of the right circular cone, when the vertex portion is cut off by a plane perpendicular to the altitude of the cone.

CHAPTER 14

1. This is from the official press release of the Nobel Assembly at Karolinska Institutet: https://www.nobelprize.org/prizes/medicine/2017/press-release/.

2. R. E. Langer, "Rene Descartes," *The American Mathematical Monthly* Vol. 44, No. 8 (October, 1937): pp. 495–512. The quote is from page 497. See also: https://www.jstor.org/stable/2301226?seq=3#metadata_info_tab_contents.

3. R. E. Langer, "Rene Descartes," *The American Mathematical Monthly* Vol. 44, No. 8 (October, 1937), pp. 495–512. The quote is from page 498. See also: https://www.jstor.org/stable/2301226?seq=3#metadata_info_tab_contents.

4. Descartes, *Discourse on the Method* (Duke Classics, 2012), p. 34.

5. Valentine Rodger Miller, *René Descartes: Principles of Philosophy*, translated, with explanatory notes, Collection des Travaux de L'Académie Internationale D'Histoire des Sciences No 30 (Netherlands: Springer), p. xvii.

6. Rene Descartes, *Discourse on the Method*, translated by John Veitch (Cosimo, Inc., 2008), p. 15.

CHAPTER 15

1. A parliament was a provincial appellate court in the Ancien Régime of France. In 1789, France had thirteen parliaments, the most important of which was the Parliament of Paris. While the English word "parliament" derives from this French term, parliaments in this sense were not legislative bodies. They consisted of a dozen or more appellate judges, or about 1,100 judges nationwide (see *Wikipedia*, s.v. "Parliament," last edited February 2, 2019, https://en.wikipedia.org/wiki/Parliament).

2. André Weil, *Zahlentheorie: Ein Gang durch die Geschichte von Hammurapi bis Legendre* (Basel, Switzerland: Birkhäuser, 1992), p. 40.

3. Michael Sean Mahoney, *The Mathematical Career of Pierre de Fermat, 1601–1665*, Second Edition (Princeton, NJ: Princeton University Press, 2018), p. 192.

4. George F. Simmons, *Calculus Gems: Brief Lives and Memorable Mathematics* (Washington, DC: Mathematical Association of America, 2007), p. 98.

5. Michael Sean Mahoney, *The Mathematical Career of Pierre de Fermat, 1601–1665*, 2nd ed. (Princeton, NJ: Princeton University Press, 2018), p. 61.

6. The method of infinite descent is a special variant of a proof by contradiction. To prove that a problem has no solution, one may be able to show that if a solution—which was in some sense related to one or more natural numbers—would exist, this would necessarily imply that another solution related to smaller natural numbers existed. The existence of this solution would then automatically imply the existence of another solution, related to even smaller natural numbers, and so forth. Since there cannot be an infinite sequence of smaller and smaller natural numbers (sooner or later one would encounter the smallest natural number with the desired property), the premise that the problem has a solution must be wrong. For example, to show that $\sqrt{2}$ is not a rational number, we may start a proof by contradiction by first assuming that $\sqrt{2}$ is rational. Then we would be able to write

$$\sqrt{2} = \frac{p}{q}$$

with p and q some natural numbers. We would then have $2q^2 = p^2$, implying that p^2 is even and thus p must be even as well (if p were odd, than p^2 cannot be even). Now, if p is even, we can write $p = 2k$ for some natural number k, which upon inserting in the last equation yields $2q^2 = 4k^2$, that is, $q^2 = 2k^2$ and therefore q must also be even. Thus, both p and q must be divisible by 2. This means that if $\sqrt{2}$ had a representation as a rational number $\frac{p}{q}$ this fraction could always be reduced by dividing p and q by 2. But this is impossible since we cannot reduce a fraction further and further, without end. This contradiction tells us that $\sqrt{2}$ cannot be rational.

7. Reinhard Laubenbacher and David Pengelley, *Mathematical Expeditions: Chronicles by the Explorers* (Springer Science & Business Media, 2013), p. 165.

8. John Tabak, *Probability and Statistics: The Science of Uncertainty*, The History of Mathematics Series (Infobase Publishing, 2014), p. 27.

9. Simon Singh, *Fermat's Last Theorem* (Fourth Estate, 1997).

CHAPTER 18

1. A vacuum tube is a device that controls electric currents between electrodes in an evacuated container. Invented in 1904, vacuum tubes were a basic component for electronics throughout the first half of the twentieth century, which saw the diffusion of radio, television, large telephone networks, as well as analog and digital computers.

2. *Wikipedia*, s.v. "ENIAC," last edited February 14, 2019, https://en.wikipedia.org/wiki/ENIAC.

3. Caren L. Diefenderfer and Roger B. Nelsen, *The Calculus Collection: A Resource for AP and Beyond* (MAA, 2019).

4. Richard T. W. Arthur, "The Remarkable Fecundity of Leibniz's Work on Infinite Series," *Annals of Science* Vol. 63, Issue 2 (2006).

5. G. W. Leibniz, *Interrelations between Mathematics and Philosophy*, edited by Norma B. Goethe, Philip Beeley, and David Rabouin (Springer, 2015), p. 146.

6. The translation of Leibniz's text can be found on the website "Leibniz Translations" by Lloyd Strickland; here is the link to article about binary arithmetic: http://www.leibniz-translations.com/binary.htm.

7. The translation of Leibniz's text can be found on the website "Leibniz Translations" by Lloyd Strickland; here is the link to article about binary arithmetic: http://www.leibniz-translations.com/binary.htm.

8. Richard C. Brown, *The Tangled Origins of the Leibnizian Calculus: A Case Study of a Mathematical Revolution* (World Scientific, 2012), p. 229.

CHAPTER 19

1. The text that follows is derived from the appendix of Alfred S. Posamentier, Robert Geretschläger, Charles Li, and Christian Spreitzer, *The Joy of Mathematics: Marvels, Novelties, and Neglected Gems That Are Rarely Taught in Math Class* (Amherst, NY: Prometheus Books, 2017), pp. 289–91.

2. This section is derived from Alfred S. Posamentier and Ingmar Lehmann, *The Secrets of Triangles: A Mathematical Journey* (Amherst, NY: Prometheus Books, 2012), p. 45.

3. This is a "biconditional" statement that indicates that if the lines are concurrent, then the equation is true; and if the equation is true, then the lines are concurrent.

4. The following section is derived from Posamentier and Lehmann, *Secrets of Triangles*, pp. 135–36 and 342.

CHAPTER 20

1. One such example is Alfred S. Posamentier and Robert L. Bannister, *Geometry: Its Elements and Structure*, 2nd ed. (New York: Dover, 2014).

2. R. A. Rankin, "Robert Simson," School of Mathematics and Statistics, University of St Andrews, Scotland, http://www-history.mcs.st-andrews.ac.uk/Biographies/Simson.html.

CHAPTER 21

1. The following biographical information is derived from J. J. O'Connor and E. F. Robertson, "Christian Goldbach," August 2006, http://www-history.mcs.st-andrews.ac.uk/Biographies/Goldbach.html.

2. Tomas Oliveira e Silva, Siegfried Herzog, and Silvio Pardi, "Emperical Verification of the Even Goldbach Conjecture and computation of Prime Gaps up to $4 \cdot 10^{18}$," *Mathematics of Computation*, Vol. 83, No. 288 (July 2014): pp. 2033–60, S 0025-5718(2013)02787-1, article electronically published on November 18, 2013.

3. H. A. Helfgott, "Major Arcs for Goldbach's Theorem," *French National Centre for Scientific Research* (May 2013).

CHAPTER 22

1. *Biographical Dictionary of Mathematicians*, Vol. 1 (New York: Charles Scribner's), p. 221.

2. Dirk Jan Struik, *A Source Book in Mathematics, 1200–1800* (in the series Princeton Legacy Library) (Princeton, NJ: Princeton University Press, 2014), p. 320.

3. *Biographical Dictionary of Mathematicians*, Vol. 1 (New York: Scribner's), p. 228.

CHAPTER 23

1. The following paragraphs are derived from Alfred S. Posamentier and Christian Spreitzer, *The Mathematics of Everyday Life* (Amherst, NY: Prometheus Books, 2018), pp. 237–41.

CHAPTER 24

1. Clifford A. Pickover, *The Math Book: From Pythagoras to the 57th Dimension, 250 Milestones in the History of Mathematics*, Milestones Series (Sterling Publishing Company, Inc., 2009), p. 180.

CHAPTER 25

1. Benjamin Libet, *Mind Time: The Temporal Factor in Consciousness* (Cambridge, MA: Harvard University Press, 2009).

2. Lev Vaidman, "Quantum Theory and Determinism," *Quantum Studies: Mathematics and Foundations*, Vol. 1, Issue 1–2 (September 2014): pp. 5–38.

3. The following biographical information is derived from J. J. O'Connor and E. F. Robertson, "Pierre-Simon Laplace," January 1999, http://www-history.mcs.st -and.ac.uk/Biographies/Laplace.html.

4. This and the following biographical information is derived from *Wikipedia*, s.v. "Pierre-Simon Laplace," last edited March 15, 2019, https://en.wikipedia.org/ wiki/Pierre-Simon_Laplace.

5. Napier Shaw, *Manual of Meteorology*, Vol. 1 (Cambridge: Cambridge University Press, 2015), p. 130.

6. In fact, it is now understood that the solar system is not stable over very long periods of time. Laplace's methods were not sufficiently precise to prove stability, but they were essential steps in the development of a precise mathematical theory of celestial motion.

7. This famous quote has often been interpreted as evidence for Laplace's atheism, but this conclusion might be wrong. Physicist Stephen Hawking shared this view of Laplace, as evidenced by what he said in a 1999 public lecture: "I don't think that Laplace was claiming that God does not exist. It's just that he doesn't intervene, to break the laws of Science." (Stephen Hawking, "Does God Play Dice?" lecture, 1999, https://web.archive.org/web/20000902184353/http://www.hawking .org.uk:80/lectures/dice.html).

8. *Napoleon's Memoirs: Napoléon I, Emperor of the French. Memoirs of the history of France during the reign of Napoleon . . . 1823–1826*, Vol. I (H. Colburn and Company, 1823), p. 116.

9. Laplace wrote this in the introduction to his work *Théorie Analytique des Probabilitiés*, the quote can be found in the MacTutor History of Mathematics archive, School of Mathematics and Statistics, University of St Andrews, Scotland.

CHAPTER 26

1. The following discussion of Mascheroni constructions is derived from Alfred S. Posamentier and Robert Geretschläger, *The Circle: A Mathematical Exploration beyond the Line* (Amherst, NY: Prometheus Books, 2016), pp. 199–215.

2. For a proof of this theorem, see Alfred S. Posamentier and Charles T. Salkind, *Challenging Problems in Geometry* (New York: Dover, 1996), p. 217.

CHAPTER 27

1. The following biographical information is derived from *Wikipedia*, s.v. "Joseph-Louis Lagrange: Biography," last modified April 12, 2019, https://en.wikipedia .org/wiki/Joseph-Louis_Lagrange.

2. *Wikipedia*, s.v. "Tautochrone Curve," last edited March 15, 2019, https://en.wikipedia.org/wiki/Tautochrone_curve.

3. The three-body problem is the problem in physics of computing the trajectory of three bodies interacting with one another.

4. T. S. Blyth and E. F. Robertson, *Further Linear Algebra* (London: Springer, 2002), p. 187.

5. J. J. O'Connor and E. F. Robertson, "Joseph-Louis Lagrange," January 1999, http://www.history.mcs.st-andrews.ac.uk/Biographies/Lagrange.html.

6. *Wikipedia*, s.v. "Joseph-Louis Lagrange: Prizes and Distinctions," last modified April 12, 2019, https://en.wikipedia.org/wiki/Joseph-Louis_Lagrange.

CHAPTER 28

1. Gina Kolata, "At Last, Shout of 'Eureka!' in Age-Old Math Mystery," *New York Times*, June 24, 1993.

2. The following biographical information is derived from *Wikipedia*, s.v. "Sophie Germain," last updated April 11, 2019, https://en.wikipedia.org/wiki/Sophie_Germain.

3. J. J. O'Connor and E. F. Robertson, "Sophie Germain," December 1996, http://www-history.mcs.st-andrews.ac.uk/Biographies/Germain.html.

4. The content of this paragraph is derived from *Wikipedia*, s.v., "Sophie Germain: Later Work in Elasticity," last updated April 11, 2019, https://en.wikipedia.org/wiki/Sophie_Germain.

5. The content of this paragraph is derived from *Wikipedia*, s.v. "Sophie Germain: Final Years," in ibid.

CHAPTER 29

1. This and the following biographical information is derived from J. J. O'Connor and E. F. Robertson, "Johann Carl Friedrich Gauss," December 1996, http://www-history.mcs.st-and.ac.uk/Biographies/Gauss.html.

2. https://thatsmaths.com/2014/10/09/triangular-numbers-eyphka/.

3. https://www.scientificamerican.com/article/are-mathematicians-finall/.

4. This and the following biographical information is derived from ibid.

CHAPTER 31

1. NBIM, Norges Bank, Statistics Norway; see, for example, "Factbox: Norway's $960 Billion Sovereign Wealth Fund," Reuters, June 2, 2017, https://www.reuters.com/article/us-norway-swf-ceo-factbox/factbox-norways-960-billion-sovereign-wealth-fund-idUSKBN18T283.

2. https://worldhappiness.report/ed/2017/.

3. "Niels Henrik Abel," Norsk Biografisk Leksikon, last updated February 13, 2009, https://nbl.snl.no/Niels_Henrik_Abel.

4. Olav Arnfinn Laudal and Ragni Piene, *The Legacy of Niels Henrik Abel: The Abel Bicentennial, Oslo, 2002* (Berlin: Springer, 2013).

5. Arild Stubhaug, *Called Too Soon by Flames: Niels Henrik Abel and His Times* (Heidelberg: Springer, 2000).

6. Arild Stubhaug, *Niels Henrik Abel and his Times: Called Too Soon by Flames Afar*, translated by R. H. Daly (Springer Science & Business Media, 2013), p. 231.

7. Krishnaswami Alladi, *Ramanujan's Place in the World of Mathematics* (New Delhi: Springer, 2013), p. 83.

8. http://www.abelprize.no/c53680/artikkel/vis.html?tid=53897.

CHAPTER 32

1. The content of this paragraph is derived from *Wikipedia*, s.v., "Évariste Galois: Final Days," last updated April 25, 2019, https://en.wikipedia.org/wiki/%C3%89variste_Galois.

2. Ibid.

CHAPTER 33

1. The content of this and the following paragraphs is derived from *Wikipedia*, s.v. "James Joseph Sylvester: Biography," last updated April 11, 2019, https://en.wikipedia.org/wiki/James_Joseph_Sylvester.

2. J. D. North, "James Joseph Sylvester," *Complete Dictionary of Scientific Biography* (Charles Scribner's Sons, 2008), available through MacTutor History of Mathematics at http://www-history.mcs.st-and.ac.uk/DSB/Sylvester.pdf, citing James Joseph Sylvester, *Collected Mathematical Papers* 4, no. 53 (1888): 588.

3. http://www-history.mcs.st-andrews.ac.uk/Quotations/Sylvester.html.

CHAPTER 34

1. Lord Byron, *Childe Harold's Pilgrimage* (1812–1818), canto 3, ll. 1–2.

2. Betty Alexandra Toole, *Ada, The Enchantress of Numbers* (Mill Valley, CA: Strawberry), pp. 240–61.

CHAPTER 40

1. At these conferences the famous Fields Medals (i.e., equivalent in mathematics to the Nobel Prize) are awarded every four years to outstanding mathematicians not above age 40.

2. The curious reader is referred to https://en.wikipedia.org/wiki/Hilbert%27s_problems.

3. Hajo G. Meyer, *Tragisches Schicksal. Das deutsche Judentum und die Wirkung historischer Kräfte: Eine Übung in angewandter Geschichtsphilosophie* (Berlin: Frank & Timme, 2008), 202.

4. See http://www.storyofmathematics.com/20th_hilbert.html.

CHAPTER 41

1. Godfrey Harold Hardy, *Collected Papers of G. H. Hardy* (Oxford: Oxford University Press, 1979).

CHAPTER 43

1. Warner Bros., 2016.

2. Robert Kanigel, *The Man Who New Infinity: A Life of the Genius Ramanujan* (New York: Macmillan, 1991).

3. "Quotations by Hardy," archived from the original on July 16, 2012, accessed November 20, 2012, https://www-history.mcs.st-andrews.ac.uk/Quotations/Hardy.html.

4. G. S. Carr, *A Synopsis of Elementary Results in Pure and Applied Mathematics* (Cambridge: Cambridge University Press, 2013).

CHAPTER 44

1. William Poundstone, *Prisoner's Dilemma: John von Neumann, Game Theory, and the Puzzle of the Bomb* (New York: Anchor, 1993).

2. Claudia Dreifus, "Maria Konnikova Shows Her Cards," *New York Times*, August 10, 2018.

CHAPTER 45

1. See https://en.wikipedia.org/wiki/Hilbert%27s_program.

CHAPTER 46

1. A. M. Turing, "Intelligent Machinery" (manuscript) *(Turing Archive, 1948), 3.*

2. "von Neumann . . . firmly emphasized to me, and to others I am sure, that the fundamental conception is owing to Turing—insofar as not anticipated by Babbage, Lovelace and others," letter by Stanley Frankel to Brian Randell, 1972, quoted in Jack Copeland, *The Essential Turing* (New York: Oxford University Press, 2004), 22.

3. See A. S. Posamentier and I. Lehmann, *The Fabulous Fibonacci Numbers* (Amherst, NY: Prometheus Books, 2007).

CHAPTER 47

1. To compute the Erdős number of an author, visit https://mathscinet.ams .org/mathscinet/freeTools.html and go to the collaboration distance calculator. The tool will automatically find a path in the MathSciNet database between any two people you wish (there is a special button for selecting Paul Erdős as one end of the path).

2. According to "Facts about Erdős Numbers and the Collaboration Graph," using the Mathematical Reviews database, the next highest article count is roughly 823, see http://oakland.edu/enp/trivia/.

3. Paul Hoffman, *The Man Who Loved Only Numbers* (New York: Hyperion, 1998).

4. ln(x) is the natural logarithm of the number, x, and is its logarithm to the base of the mathematical constant e, where e is an irrational and transcendental number approximately equal to 2.718281828459 . . .

CHAPTER 48

1. A. S. Posamentier and H. A. Hauptman, *101+ Great Ideas for Introducing Key Concepts in Mathematics*, 2nd ed. (Thousand Oaks, CA: Corwin Press, 2006).

CHAPTER 49

1. Benoit B. Mandelbrot, *The Fractal Geometry of Nature* (New York: W. H. Freeman, 1983), 1.

2. The Koch Snowflake was named in 1904 after the Swedish mathematician Helge von Koch (1870–1924).

3. Named in 1915 after the Polish mathematician Waclaw Sierpiński (1882–1969).

4. The complex plane is the two-dimensional representation of complex numbers with a real axis and an imaginary axis.

5. In fact, figure 49.7 is only an approximation of the Mandelbrot set. In actuality, we cannot know for sure whether a number c lies in the Mandelbrot set, because to determine that with absolute certainty we would need to iterate the "test" an infinite number of times. But even with computers, we can obviously iterate anything only a finite number of times. But it so happens that the sequence formed by iterating the rule to a certain value of c may behave differently only after a very large number of iterations. So we can make our approximation better by iterating a great number of times. Still, this will not lead to absolute accuracy.

6. A cardioid is a heart-shaped curve generated by a fixed point on a circle as it rolls around another circle of equal radius.

7. *The Economist*, October 21, 2010.

References

Aaboe, Asger. *Episodes from the Early History of Mathematics*. New York: Random House, 1964.

Anglin, W. S., and J. Lambek. *The Heritage of Thales*. New York: Springer, 1995.

Artmann, Benno. *Euclid—The Creation of Mathematics*. New York: Springer, 1999.

Ball, W. W. Rouse. *A Short Account of the History of Mathematics*. New York: Dover, 1960.

Bell, E. T. *Men of Mathematics*. New York: Simon & Schuster, 1937.

Berlinghoff, William P., and Fernando Q. Gouvea. *Math Through the Ages: A Gentle History for Teachers and Others*. Washington, DC: Mathematical Association of America, 2004.

Boyer, Carl B. *The History of the Calculus and Its Conceptual Development*. New York: Dover, 1949.

———. *A History of Mathematics*. New York: Wiley, 1968.

Bunt, Lucas N., Philip S. Jones, and Jack D. Bedient. *The Historical Roots of Elementary Mathematics*. Englewood Cliffs, NJ: Prentice Hall, 1976.

———. *The Historical Roots of Elementary Mathematics*. New York: Dover, 1988.

Burton, David M. *The History of Mathematics: An Introduction*. Boston: Allyn and Bacon, 1985.

Cajori, Florian. *A History of Mathematical Notations*. Vol. I, II. LaSalle, IL: Open Court, 1952.

———. *A History of Mathematics*. New York: Chelsea, 1985.

Calinger, Ronald. *Classics of Mathematics*. Oak Park, IL: Moore, 1982.

Cardano, Girolamo. *ARS Magna or The Rules of Algebra*. New York: Dover, 1968.

Cohen, Patricia Cline. *A Calculating People: The Spread of Numeracy in Early America*. Chicago: University of Chicago, 1982.

Coolidge, Julian Lowell. *The Mathematics of Great Amateurs*. New York: Dover, 1963.

Dunham, William. *Euler, The Master of Us All*. Washington, DC: Mathematical Association of America, 1999.

————. *Journey Through Genius*. New York: Wiley, 1990.

Dunnington, G. Waldo. *Carl Friedrich Gauss: Titan of Science*. Washington, DC: Mathematical Association of America, 1955.

Ekeland, Ivar. *Mathematics and the Unexpected*. Chicago: University of Chicago Press, 1988.

Eves, Howard. *Great Moments in Mathematics After 1650*. Washington, DC: Mathematical Association of America, 1981.

————. *Great Moments in Mathematics Before 1650*. Washington, DC: Mathematical Association of America, 1980.

————. *An Introduction to the History of Mathematics*. Philadelphia: Saunders College Publishing, 1983.

Fauvel, John, and Jeremy Gray, eds. *The History of Mathematics: A Reader*. Milton Keynes, UK: Open University, 1987.

Friedrichs, K. O. *From Pythagoras to Einstein*. New York: L. W. Singer, 1965.

Grattan-Guinness, Ivor. *The Norton History of the Mathematical Sciences*. New York: Norton, 1998.

Gullberg, Jan. *Mathematics from the Birth of Numbers*. New York: Norton, 1997.

Hall, A. Rupert. *From Galileo to Newton*. New York: Dover, 1981.

Heath, Thomas. *A History of Greek Mathematics, Vols. I, II,* New York: Dover, 1981.

Heath, Thomas L. *A Manual of Greek Mathematics*. New York: Dover, 1963.

————. *Euclid's Elements*. Vol. 1–3. New York: Dover, 1956.

Hofmann, Joseph E. *The History of Mathematics to 1800*. Totowa, NJ: Littlefield, Adams, 1967.

Infeld, Leopold. *Whom the God's Love: The Story of Evariste Galois*. Reston, VA: National Council of Teachers of Mathematics, 1975.

James, Ioan. *Remarkable Mathematicians, from Euler to von Neumann*. Washington, DC: Mathematical Association America, 2002.

Kanigel, Robert. *The Man Who Knew Infinity: A Life of the Genius Ramanujan*. New York: Charles Scribner's, 1991.

Karpinski, Louis Charles. *The History of Arithmetic*. New York: Rand McNally, 1925.

Katz, Victor J. *A History of Mathematics: An Introduction*. 3rd ed. Boston: Addison-Wesley, 2009.

Klein, Jacob. *Greek Mathematical Thought and the Origin of Algebra*. New York: Dover, 1968.

Krantz, Steven G. *Mathematical Apocrypha: Stories and Anecdotes of Mathematicians and the Mathematical*. Washington, DC: Mathematical Association of America, 2002.

Lasserre, Francois. *The Birth of Mathematics in the Age of Plato*. Larchmont, NY: American Research Council, 1964.

Lewinter, Marty, and William Widulski. *The Saga of Mathematics: A Brief History*. Upper Saddle River, NJ: Prentice Hall, 2002.

Mankiewicz, Richard. *The Story of Mathematics*. Princeton, NJ: Princeton University Press, 2000.

Maor, Eli. *To Infinity and Beyond: A Cultural History of the Infinite*. Princeton, NJ: Princeton University Press, 1991.

Meschkowski, Herbert. *Ways of Thought of Great Mathematicians*. San Francisco: Holden-Day, 1964.

Moritz, Robert Edouard. *On Mathematics: A Collection of Witty, Profound, Amusing Passages About Mathematics and Mathematicians*. New York: Dover, 1942.

Morrison, Philip, and Emily Morrison. *Charles Babbage: On the Principles and Development of the Calculator and Other Seminal Writings by Charles Babbage and Others*. New York: Dover, 1961.

Ore, Oystein. *Cargano the Gambling Scholar*. New York: Dover, 1953.

Perl, Teri. *Math Equals: Biographies of Women Mathematicians + Related Activities*. Menlo Park, CA: Addison-Wesley, 1978.

Petronis, Vida, ed. *Biographical Dictionary of Mathematicians*. Vol. 1–4. New York: Charles Scribner's, 1991.

Phillips, Esther R. *Studies in the History of Mathematics*. Washington, DC: Mathematical Association of America, 1987.

Posamentier, Alfred S. and Ingmar Lehmann. *Pi: A Biography of the World's Most Mysterious Number*. Amherst, NY: Prometheus Books, 2007.

Posamentier, Alfred S. and Ingmar Lehmann. *The (Fabulous) Fibonacci Numbers*. Amherst, NY: Prometheus Books, 2004.

Reid, Constance. *Hilbert*. New York: Springer, 1970.

Robins, Gay, and Charles Shute. *The Rhind Mathematical Papyrus: An Ancient Egyptian Text*. New York: Dover, 1987.

Rowe, David E., and John McCleary, eds. *The History of Modern Mathematics*. Vol. I, *Ideas and Their Reception*. Boston: Academic Press, 1989.

———. *The History of Modern Mathematics*. Vol. II, *Institutions and Applications*. Boston: Academic Press, 1989.

Rudman, Peter S. *How Mathematics Happen: The First 50,000 Years*. Amherst, NY: Prometheus, 2007.

Sanford, Vera. *A Short History of Mathematics*. Boston: Houghton Mifflin, 1958.

Smith, David Eugene. *History of Mathematics*. Vol. I and II. New York: Dover, 1951.

———. *A Source Book in Mathematics*. New York: McGraw-Hill, 1929.

Smith, David Eugene, and Marcia L. Latham. *The Geometry of René Descartes*. New York: Dover, 1925.

Stein, Sherman. *Archimedes: What Did He Do Besides Cry Eureka?* Washington, DC: Mathematical Association of America, 1999.

Stillwell, John. *Mathematics and Its History*. New York: Springer, 1989.

Struik, Dirk J. *A Concise History of Mathematics*. New York: Dover, 1948.

————. *A Source Book in Mathematics, 1200–1800*. Princeton, NJ: Princeton University, 1986.

Turnbull, Herbert Westren. *The Great Mathematicians*. New York: New York University, 1961.

Van der Waerden, B. L. *Geometry and Algebra in Ancient Civilizations*. New York: Springer, 1983.

————. *A History of Algebra: From al-Khwarizmi to Emmy Noether*. New York: Springer, 1980.

————. *Science Awakening: Egyptian, Babylonian and Greek Mathematics*. New York: Wiley, 1963.